Springer Theses

Recognizing Outstanding Ph.D. Research

More information about this series at http://www.springer.com/series/8790

Aims and Scope

The series "Springer Theses" brings together a selection of the very best Ph.D. theses from around the world and across the physical sciences. Nominated and endorsed by two recognized specialists, each published volume has been selected for its scientific excellence and the high impact of its contents for the pertinent field of research. For greater accessibility to non-specialists, the published versions include an extended introduction, as well as a foreword by the student's supervisor explaining the special relevance of the work for the field. As a whole, the series will provide a valuable resource both for newcomers to the research fields described, and for other scientists seeking detailed background information on special questions. Finally, it provides an accredited documentation of the valuable contributions made by today's younger generation of scientists.

Theses are accepted into the series by invited nomination only and must fulfill all of the following criteria

- They must be written in good English.
- The topic should fall within the confines of Chemistry, Physics, Earth Sciences, Engineering and related interdisciplinary fields such as Materials, Nanoscience, Chemical Engineering, Complex Systems and Biophysics.
- The work reported in the thesis must represent a significant scientific advance.
- If the thesis includes previously published material, permission to reproduce this must be gained from the respective copyright holder.
- They must have been examined and passed during the 12 months prior to nomination.
- Each thesis should include a foreword by the supervisor outlining the significance of its content.
- The theses should have a clearly defined structure including an introduction accessible to scientists not expert in that particular field.

Colin Howard

Measuring, Interpreting and Translating Electron Quasiparticle – Phonon Interactions on the Surfaces of the Topological Insulators Bismuth Selenide and Bismuth Telluride

Doctoral Thesis accepted by Boston University, Boston, Massachusetts, USA

Colin Howard
Boston University
Boston, MA, USA

ISSN 2190-5053   ISSN 2190-5061 (electronic)
Springer Theses
ISBN 978-3-319-44722-3   ISBN 978-3-319-44723-0 (eBook)
DOI 10.1007/978-3-319-44723-0

Library of Congress Control Number: 2016950525

© Springer International Publishing Switzerland 2016
This work is subject to copyright. All rights are reserved by the Publisher, whether the whole or part of the material is concerned, specifically the rights of translation, reprinting, reuse of illustrations, recitation, broadcasting, reproduction on microfilms or in any other physical way, and transmission or information storage and retrieval, electronic adaptation, computer software, or by similar or dissimilar methodology now known or hereafter developed.
The use of general descriptive names, registered names, trademarks, service marks, etc. in this publication does not imply, even in the absence of a specific statement, that such names are exempt from the relevant protective laws and regulations and therefore free for general use.
The publisher, the authors and the editors are safe to assume that the advice and information in this book are believed to be true and accurate at the date of publication. Neither the publisher nor the authors or the editors give a warranty, express or implied, with respect to the material contained herein or for any errors or omissions that may have been made.

Printed on acid-free paper

This Springer imprint is published by Springer Nature
The registered company is Springer International Publishing AG
The registered company address is: Gewerbestrasse 11, 6330 Cham, Switzerland

*For my parents and María, without whom
I would not have been able to do this*

Supervisor's Foreword

The new class of materials coined topological insulators (TIs) has been the subject of extensive studies over the past 5 years and continues to be one of the most active research areas of condensed matter physics. The main attraction in studying these materials, from both fundamental and technological perspectives, has been the presence of chiral Dirac fermion quasiparticles (DFQs) that define a robust metallic surface state protected against backscattering by time-reversal symmetry. Strong spin-orbit interactions lock the Dirac fermion quasiparticles' (DFQs) spin to the state wave vector in a mutually perpendicular configuration, giving the Dirac cone a definite chirality. Consequently, DFQs on the surface cannot backscatter from lattice vacancies, grain boundaries, phonons, etc. into their time-reversed counterparts. Technical improvements may minimize defects, but phonons are always present. Despite these constraints, it was recently found that the DFQs strongly interact with surface boson excitations, especially phonon and coupled plasmon-spin excitations. Consequently, DFQ-phonon interaction should be a dominant scattering mechanism for Dirac fermions on these surfaces at finite temperatures. Hence, the electron–phonon interaction is of exceptional importance when assessing the feasibility of promising applications in technologies such as spintronics and quantum computing.

Most of the reported studies adopted a noninteracting, single-particle approach to the DFQ system. However, recently, there have been several experimental and theoretical reports on the interaction of the DFQs with surface phonons. Prior to Dr. Howard's thesis work, there has been little consensus about the magnitude of this coupling as evidenced by widely varying values of the electron–phonon coupling parameter λ appearing in the literature.

Dr. Howard's thesis presents experimental and theoretical results about the surface dynamics and the surface Dirac fermion (DF) spectral function of strong topological insulators Bi_2Te_3, and Bi_2Se_3, and describes the corresponding techniques used. The experimental results, employing inelastic helium atom-surface scattering techniques, reveal the presence of a prominent Kohn anomaly in the measured surface phonon dispersion of a low-lying optical mode and the absence of surface Rayleigh acoustic phonons. With the aid of fitting the experimental data to theoretical models employing phonon Matsubara functions, he was able to extract the matrix elements of the coupling Hamiltonian and to determine the modifications to the surface phonon propagator encoded in the phonon self-energy. This, in turn, allowed him to calculate, for the first time, the phonon mode-specific electron–phonon coupling $\lambda(\mathbf{q})$ from experimental data and to obtain an average coupling significantly higher than typical values obtained for metals, underscoring the strong interaction between optical surface phonons and surface Dirac fermions in topological insulators. Finally, to connect to experimental λ values obtained from photoemission spectroscopy, he constructed an electronic DFQ Matsubara function using the determined electron–phonon matrix elements and the optical phonon dispersion. Dr. Howard was then able to extract the DFQ spectral function and the density of states for comparison with angle-resolved photoemission (ARPES) and scanning tunneling spectroscopy results. The ensuing spectral function revealed several important features. First, the footprints of phonon interactions occur on an energy scale of the order of 1 meV, which sets a necessary energy resolution of that magnitude or better to observe these features and extract a reliable value for λ. Second, he found that the electron–phonon coupling parameter extracted from the spectral function was strongly temperature dependent, which invalidates extraction methodologies that assume a temperature-independent λ and a linear temperature dependence of the corresponding spectral linewidth.

I hope that the methodology and techniques presented in Dr. Howard's thesis could hold promise for determining how EPC, as well as other quasiparticle interactions, modifies the electronic structure in a variety of condensed matter systems. I also hope that the entire work would be useful for students and researchers.

Boston, MA, USA Michael El-Batanouny
May 2016

Abstract

The following dissertation presents a comprehensive study of the interaction between Dirac fermion quasiparticles (DFQs) and surface phonons on the surfaces of the topological insulators Bi_2Se_3 and Bi_2Te_3. Inelastic helium atom surface scattering (HASS) spectroscopy and time of flight (TOF) techniques were used to measure the surface phonon dispersion of these materials along the two high-symmetry directions of the surface Brillouin zone (SBZ). Two anomalies common to both materials are exhibited in the experimental data. First, there is an absence of Rayleigh acoustic waves on the surface of these materials, pointing to weak coupling between the surface charge density and the surface acoustic phonon modes and potential applications for soundproofing technologies. Secondly, both materials exhibit an out-of-plane polarized optical phonon mode beginning at the SBZ center and dispersing to lower energy with increasing wave vector along both high-symmetry directions of the SBZ. This trend terminates in a V-shaped minimum at a wave vector corresponding to $2k_F$ for each material, after which the dispersion resumes its upward trend. This phenomenon constitutes a strong Kohn anomaly and can be attributed to the interaction between the surface phonons and DFQs.

To quantify the coupling between the optical phonons experiencing strong renormalization and the DFQs at the surface, a phenomenological model was constructed based within the random phase approximation. Fitting the theoretical model to the experimental data allowed for the extraction of the matrix elements of the coupling Hamiltonian and the modifications to the surface phonon propagator encoded in the phonon self energy. This allowed, for the first time, calculation of phonon mode-specific quasiparticle-phonon coupling $\lambda_\nu(\mathbf{q})$ from experimental data. Additionally, an averaged coupling parameter was determined for both materials yielding $\bar{\lambda}^{Te} \approx 2$ and $\bar{\lambda}^{Se} \approx 0.7$. These values are significantly higher than those of typical metals, underscoring the strong coupling between optical surface phonons and DFQs in topological insulators.

In an effort to connect experimental results obtained from phonon and photoemission spectroscopies, a computational process for taking coupling information from the phonon perspective and translating it to the DFQ perspective was derived. The procedure involves using information obtained from HASS measurements (namely the coupling matrix elements and optical phonon dispersion) as input to a Matsubara Green function formalism, from which one can obtain the real and imaginary parts of the DFQ self energy. With these at hand it is possible to calculate the DFQ spectral function and density of states, allowing for comparison with photoemission and scanning tunneling spectroscopies. The results set the necessary energy resolution and extraction methodology for calculating $\bar{\lambda}$ from the DFQ perspective. Additionally, determining $\bar{\lambda}$ from the calculated spectral functions yields results identical to those obtained from HASS, proving the self-consistency of the approach.

Acknowledgments

First and foremost, I would like to thank the most important person in my life, María Jorgelina González, who was crazy enough to become my wife during this whole ordeal. Without her, this dissertation would definitely not have been written. A constant source of inspiration and support, I will never be able to thank her enough nor express how important she is for my life. That being said, I can still try. So, María, I offer you my deepest and most sincere thanks for being the perfect companion over these past 5 years. I cannot wait to see where our journey takes us.

Second, I would like to thank my parents, Gail, Judd and Lori. Ever since I was young, both my father and mother instilled a great wonderment of the unknown in me. I feel that my passion for physics and desire to delve into the frontiers of knowledge are at least partly due to them. They also taught me the value of hard and honest work, whether it be physical or mental, and for that lesson, I am eternally grateful. It sometimes pains me to think about the sacrifices they all made in their own lives so that I had the opportunity at an education. I hope that this dissertation will, in some small way, show that those sacrifices were not in vain.

Of course, a proper acknowledgment would not be complete without thanking those in the Boston University physics department who touched my life, through physics or otherwise. I would like to thank Mirtha Cabello, for always being so helpful, practicing Spanish with me, and helping María and I find a place to live upon moving to Boston. I would also like to sincerely thank Sidney Redner for being perhaps the most lucid and kind professor of my graduate school tenure. Finally, I am extremely grateful to Claudio Chamon. My respect for him stems not only from his teaching ability but also his humility and profound guidance on my own research.

Finally, I owe an enormous debt of gratitude to my advisor and mentor Michael El-Batanouny. I often smile thinking that much of what I know about the finer details of our physical world is due to a man who started his own journey halfway across the world in a completely different culture. He has accomplished so much in his own life that it is often difficult to imagine where he found time to do it all, but I look at that as a source of inspiration rather than intimidation. I thank him for his endless patience, profound insight, and occasional history lesson! I thank him for the thorough conversations, long experiments, and countless Skype calls. Perhaps, they were not all enjoyable at the time, but they have made me the scientist that I am today, and for that, I am forever grateful.

Contents

1 **Introduction** 1
   References 5

2 **Properties of Bi_2Se_3 and Bi_2Te_3** 7
   2.1   Crystal Structure 7
   2.2   Bulk Vibrational Structure 9
   2.3   Electronic Structure 11
   References 14

3 **Helium Atom-Surface Scattering (HASS)** 15
   3.1   The Benefits of HASS 15
   3.2   The Surface Interaction Potential 16
   3.3   The Kinematics of HASS 18
       3.3.1   Elastic Scattering 18
       3.3.2   Inelastic Scattering and Time-of-Flight Technique 19
   References 22

4 **Experimental Apparatus and Technique** 23
   4.1   Surface Laboratory Facilities 23
   4.2   Source Chamber 23
   4.3   Target Chamber 26
       4.3.1   Production and Monitoring of UHV 27
       4.3.2   Sample Manipulator 27
       4.3.3   Sample Cleaver 28
       4.3.4   Helium Detector 29
   References 31

5 **Pseudocharge Phonon Model** 33
   5.1   Fundamentals of the Model 33
   5.2   Adiabatic Approximation, Ionic Self-Terms, and PC Self-Terms 34
   5.3   Bulk Parameters 36
   5.4   Surface Parameters 41
   References 42

6 HASS Results from the Surface of Bi_2Se_3 and Bi_2Te_3 ... 43
6.1 Elastic and Inelastic Scattering Results ... 43
6.2 Calculation of EPC Parameter in the Random Phase Approximation ... 48
References ... 53

7 Translating Between Electron and Phonon Perspectives ... 55
7.1 Motivation ... 55
7.2 DFQ Self-Energy Formalism ... 56
7.3 Computational Results ... 58
7.4 Additional Supporting Results ... 62
References ... 64

8 Conclusion and Future Directions ... 65
8.1 Summary ... 65
8.2 Future Work ... 66
References ... 70

A Supplemental Material for Electron Self-Energy Analysis ... 71
A.1 Electron Green's Function ... 71
A.2 Bosonic Sums ... 77

B Numerical Evaluation of the DFQ Self-Energy ... 81
B.1 Hole Term ... 82
    B.1.1 Above Dirac Point ... 82
    B.1.2 Below Dirac Point ... 86
B.2 Particle Term ... 86
    B.2.1 Above Dirac Point ... 86
    B.2.2 Below Dirac Point ... 87
B.3 Interband Transitions ... 88

List of Abbreviations

2D	Two dimensional
ARPES	Angle-resolved photoemission spectroscopy
BZ	Brillouin zone
DFQ	Dirac fermion quasiparticle
EPC	Electron-phonon coupling
FPGA	Field programmable gate array
FWHM	Full width at half maximum
HASS	Helium atom-surface scattering
MCPEM	Multichannel plate electron multiplier
QHE	Quantum hall effect
QSHE	Quantum spin hall effect
RGA	Residual gas analyzer
SBZ	Surface Brillouin zone
TCI	Topological crystalline insulator
TI	Topological insulator
TOF	Time of flight
TRS	Time-reversal symmetry
UHV	Ultra-high vacuum

Chapter 1
Introduction

The study of topological order, particularly in the context of the band theory of solids, is a blossoming field that has returned to the forefront of condensed matter physics within the past 10 years. Several fascinating classes of recently discovered topological materials, including topological insulators (TIs) and topological crystalline insulators (TCIs), display very rich physics. These materials are host to topologically protected metallic surface states that are manifest as chiral Dirac fermion quasiparticles. As such, the surfaces of these crystals have and continue to be fruitful environments for studying a variety of interesting phenomena including axion dynamics, proximity induced superconductivity, and Majorana fermions. In addition to their intellectual appeal, they also hold promise as candidates for use in fledgling technologies such as spin-electronics (spintronics) and quantum computing. The protected metallic surface states have many favorable characteristics including a definite helicity and robustness in the face of non-magnetic disorder, which could allow industry to overcome existing limitations in electronic miniaturization resulting from dissipation in electronic transport.

For much of the twentieth century ordered phases of matter were understood using Landau's approach [1] in which any particular phase may be characterized by the symmetry it breaks. For example, crystals break continuous translational symmetry as well as certain types of rotational symmetry in position space. Similarly, the field produced by a ferromagnet breaks both rotational symmetry in spin space and time reversal invariance. However, beginning in the early 1980s with the discovery of both the integer and fractional quantum Hall effect (QHE) [2, 3], a new type of order rooted in topology rather than symmetry began to emerge. The near perfect quantization of the transverse Hall conductivity $\sigma_{xy} = ne^2/\hbar$ (with n integer) across samples of varying composition heralded the onset of the first observed topological insulating phase of matter. Two years later Thouless, Kohmoto, Nightingale, and den Nijs showed [4] that this phase could be characterized by a

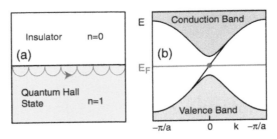

Fig. 1.1 (a) Semi-classical picture of the edge modes in the integer QHE originating from skipping Landau orbitals at the edge of the material. (b) Depiction of the electronic band structure in the integer QHE showing the branch formed by the surface modes that occupies the bulk band gap. The Fermi energy must always cross this branch regardless of its position in the bulk band gap. Figure from [6]

new invariant integer topological quantum number n dubbed the TKNN invariant or first Chern number, which, in this case, is the very coefficient appearing in the Hall conductivity.

The topology of the situation is more apparent when we consider the fact that n is a bona-fide indicator of the topological class of the insulating band structure; members of different classes cannot be adiabatically converted into one another without closing the electronic gap at the interface between the two. In the context of the QHE we have a material in an insulating phase described by finite n embedded in vacuum, which can be viewed as a type of non-topological (i.e., $n = 0$) insulator with a gap separating particle excitations from the negative energy Dirac sea. Thus, at the boundary of these two regions the gap must close to usher the change in n. Indeed, one finds spin-polarized edge modes [5] that occupy the bulk band gap. Semi-classically, these edge modes can be understood in terms of the circular electron Landau orbits skipping off the repulsive potential at the edge of the material (see Fig. 1.1a). However, because of the periodic crystal potential, these edge modes will disperse with crystal momentum **k** instead of forming a flat Landau level. The edge modes form a branch of the electronic dispersion that connects the valence band maximum to the conduction band minimum and thus crosses the Fermi level at a single point as shown in Fig. 1.1b. In such a scenario the chemical potential must always intersect at least one electronic state regardless of doping or gating, giving the edge of the material a metallic character. The result that unique boundary states exist at the interface between topologically distinct regions had been noted even before [7, 8] the discovery of the integer QHE. The phenomenon is so ubiquitous that it is now known as the *bulk-boundary correspondence.*

After the discovery of the integer and fractional QHE and the ensuing edge modes, theorists sought models that would exhibit protected boundary modes without breaking time reversal symmetry (recall that the quantum Hall physics requires the presence of a large perpendicular magnetic field). Indeed, as we will see, topological band theory is one of the few examples in physics where theory has driven experimentation and not vice versa. These studies resulted in the prediction

1 Introduction

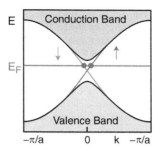

Fig. 1.2 Idealized band structure of the QSHE. Unlike the QHE the Fermi level crosses two electronic states with opposing group velocity and spin orientation. Figure from [6]

of the quantum spin Hall effect (QSHE) [9–13], a phenomenon that can be thought of as two identical copies of the integer QHE superimposed. In this scenario the edge modes are not unidirectional; degenerate left-moving and right-moving modes with opposite spin polarization coexist (see Fig. 1.2), constituting sets of Kramers pairs. Although there is no net flow of charge, there is a net spin current. Moreover, since oppositely propagating modes have opposite spins, backscattering is suppressed in the presence of non-magnetic perturbations, making the edge states quite robust (the reader may begin to understand why such materials exhibit favorable characteristics for use in spintronic devices). Although the QSHE state has a trivial TKNN invariant, it can be characterized by a different topological quantum number known as the $\mathbb{Z}_2$ invariant ν. This invariant takes integer values 0 or 1 (mod 2), corresponding to a trivial (no edge states) or topological (possessing edge states) phase, respectively. Theoretical predictions of the QSHE were eventually substantiated by measurements performed on 2D HgTe quantum well structures [14] showing a quantized residual conductance from these edge states when samples were driven into an insulating regime via a variable gate voltage.

Experimental confirmation of the QSHE ushered theorists in the condensed matter physics community to generalize these ideas to three-dimensional materials. In this case the boundary between the sample and continuum is not an *edge* but rather a *surface*. As such, the search began for time reversal invariant Hamiltonians possessing protected surface states, yielding fruitful results. A full topological description of this newly predicted phase, proposed by three independent groups [15–17], necessitated a set of four $\mathbb{Z}_2$ invariants denoted (ν_0; ν_1,ν_2,ν_3), rather than the single integer quantum number of the QHE and QSHE. It should be noted that these three-dimensional *topological insulators* can be divided immediately into two groups according to the first $\mathbb{Z}_2$ invariant. Those with $\nu_0 = 0$ are known as *weak* topological insulators and can be thought of as merely as stacked copies of the QSHE. In this scenario the surface states only appear on certain surfaces of the bulk crystal, propagate anisotropically, and are subject to localization and hybridization. Those with $\nu_0 = 1$ belong to a new class of matter not derivable directly from the QSHE and are known as *strong* topological insulators, whose robust surface states appear on any crystal termination. Much like the QHE and IQHE the surface states are manifest as linearly dispersing electronic states that span the bulk band gap. In fact, one can get a rough picture of the surface dispersion by taking the electronic

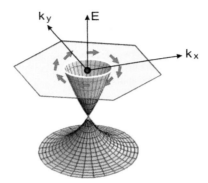

Fig. 1.3 Idealized band structure of a strong 3D topological insulator (in this case Bi_2Se_3) whose Fermi surface encloses a single Dirac point. The overlay demonstrates the orientation of the surface electronic structure within the SBZ. Figure from [6]

structure presented in Fig. 1.2 and revolving it about the vertical axis lying in the page and passing through $\mathbf{k} = 0$. The resulting conical shapes form what is known as a *Dirac cone* with the point where they touch being known as the *Dirac point*. An example of the Dirac cone band structure is shown in Fig. 1.3. Generally speaking it is possible to have multiple Dirac cones in the surface Brillouin zone (SBZ) centered about any time reversal invariant momenta. However, the Fermi surface of a strong topological insulator must enclose an *odd* number of Dirac points and thus cross an *odd* number of electronic states along a particular high-symmetry direction. Again, this makes it impossible to "gap out" such states by tuning the chemical potential, much like was shown previously for the IQHE and QSHE.

Generally speaking, determining the entire set of $\mathbb{Z}_2$ invariants for a particular material requires careful evaluation of the products of matrix elements of the time reversal operator evaluated at the time reversal invariant momenta of the bulk Brillouin zone (BZ). However, Fu and Kane showed [18] that the ordeal could be simplified considerably for materials possessing inversion symmetry, where the parity of individual electronic bands is conserved. In particular they proved that the resulting sign of the product of parities (even corresponding to $+1$ and odd corresponding to -1) of all occupied bands was sufficient to determine whether the material was in a topologically unique phase. Insulators with a positive product of parities are said to be trivial and equivalent to vacuum while those where the product is odd are topological. In addition, the insight of Fu and Kane drove the search for systems exhibiting band inversion between valence and conduction bands of opposite parity. Since an insulator without any band inversion is guaranteed to be trivial, one can be sure that an inversion between bands of opposite parity will change the sign of the product of parities and thus drive the material into the topological phase. Thus began the search for narrow-gap semiconductors exhibiting strong spin-orbit coupling capable of driving band inversion while simultaneously preserving time reversal invariance. It was not long before the first three-dimensional TIs were realized in the prototypical materials $Bi_{1-x}Sb_x$, Bi_2Se_3, Bi_2Te_3, and Sb_2Te_3; the predicted metallic surface states protected by time reversal symmetry (TRS) were clearly observed in ARPES measurements [19–26].

With the successful prediction and observation of protected metallic states on the surfaces of topological insulators, physicists began to wonder if such protected states could be derived from a crystal symmetry, rather than TRS. Eventually it was proved [27] that mirror symmetry in the surface plane could provide the necessary protection and also lead to electronic states occupying the bulk band gap. These new materials, dubbed topological *crystalline* insulators, are heavy IV-VI semiconductors that crystalize into the rocksalt structure, including the alloys $Pb_{1-x}Sn_xTe$ and $Pb_{1-x}Sn_xSe$. Unlike conventional TIs, TCIs may have a topologically trivial (even) $\mathbb{Z}_2$ while still having protected surface states owing to the presence of the mirror planes. As such, their interesting properties are encoded by a different topological quantity, the mirror Chern number. It has recently been shown [28] that the four Dirac cones within the first SBZ possess pronounced spin texture owing to spin-orbit coupling, much like nominal TIs. The discovery of the role discrete symmetries can play in modifying the band structure of these topological materials is fascinating from a fundamental physics perspective and also holds promise for emerging technology.

As mentioned before, there is speculation that the protected chiral states on the surfaces of both TIs and TCIs could be used for applications in the fields of quantum computing and spintronics. However, in order for these electronic states to be truly useful, one needs to quantify just how "protected" they are. By this I mean that, although the spin-texture of the Fermi surfaces of these materials prevents backscattering (and thus localization) from non-magnetic perturbations, many other scattering pathways still exist whereby the electrons crystal momentum is not totally reversed but still redirected. Such scattering implies a coupling between the DFQs and their environment which can lead to the loss of phase information, which could limit the usefulness of these states, especially for quantum computing. Phonons, defects, grain boundaries, etc. can all contribute to said scattering. Whereas the refinement of growth techniques may minimize crystal imperfections, phonons will always be present at finite temperatures. Therefore, it is imperative that one understands the electron–phonon interaction in these materials in order to assess their viability for technological applications. In this dissertation I will present a comprehensive study of electron–phonon interaction in the strong 3D topological insulators Bi_2Se_3 and Bi_2Te_3 using helium atom surface scattering (HASS) spectroscopy.

References

1. L. Landau, On the theory of phase transitions, in *em collected papers of L.D. Landau*, ed. by D. Ter Haar (Pergamon, London, 1965), pp. 193–216
2. K.V. Klitzing, G. Dorda, M. Pepper, New method for high-accuracy determination of the fine-structure constant based on quantized hall resistance. Phys. Rev. Lett. **45**, 494–497 (1980)
3. D.C. Tsui, H.L. Stormer, A.C. Gossard, Two-dimensional magnetotransport in the extreme quantum limit. Phys. Rev. Lett. **48**, 1559–1562 (1982)
4. D.J. Thouless, M. Kohmoto, M.P. Nightingale, M. den Nijs, Quantized hall conductance in a two-dimensional periodic potential. Phys. Rev. Lett. **49**, 405–408 (1982)

5. B.I. Halperin, Quantized hall conductance, current-carrying edge states, and the existence of extended states in a two-dimensional disordered potential. Phys. Rev. B **25**, 2185–2190 (1982)
6. M.Z. Hasan, C.L. Kane, Colloquium. Rev. Mod. Phys. **82**, 3045–3067 (2010)
7. R. Jackiw, C. Rebbi, Solitons with fermion number 1/2. Phys. Rev. D **13**, 3398–3409 (1976)
8. W.P. Su, J.R. Schrieffer, A.J. Heeger, Solitons in polyacetylene. Phys. Rev. Lett. **42**, 1698–1701 (1979)
9. S. Murakami, N. Nagaosa, S.-C. Zhang, Dissipationless quantum spin current at room temperature. Science **301**(5638), 1348–1351 (2003)
10. S. Murakami, Quantum spin hall effect and enhanced magnetic response by spin-orbit coupling. Phys. Rev. Lett. **97**, 236805 (2006)
11. C.L. Kane, E.J. Mele, Quantum spin hall effect in graphene. Phys. Rev. Lett. **95**, 226801 (2005)
12. B.A. Bernevig, S.-C. Zhang, Quantum spin hall effect. Phys. Rev. Lett. **96**, 106802 (2006)
13. B.A. Bernevig, T.L. Hughes, S.-C. Zhang, Quantum spin hall effect and topological phase transition in HgTe quantum wells. Science **314**(5806), 1757–1761 (2006)
14. M. König, S. Wiedmann, C. Brüne, A. Roth, H. Buhmann, L.W. Molenkamp, X.-L. Qi, S.-C. Zhang, Quantum spin hall insulator state in HgTe quantum wells. Science **318**(5851), 766–770 (2007)
15. L. Fu, C.L. Kane, E.J. Mele, Topological insulators in three dimensions. Phys. Rev. Lett. **98**, 106803 (2007)
16. J.E. Moore, L. Balents, Topological invariants of time-reversal-invariant band structures. Phys. Rev. B **75**, 121306 (2007)
17. R. Roy, Topological phases and the quantum spin hall effect in three dimensions. Phys. Rev. B **79**, 195322 (2009)
18. L. Fu, C.L. Kane, Topological insulators with inversion symmetry. Phys. Rev. B **76**, 045302 (2007)
19. Y. Xia, D. Qian, D. Hsieh, L. Wray, A. Pal, H. Lin, A. Bansil, D. Grauer, Y.S. Hor, R.J. Cava, M.Z. Hasan, Observation of a large-gap topological-insulator class with a single Dirac cone on the surface. Nat. Phys. **5**(6), 398–402, 06 (2009)
20. Y.L. Chen, J.G. Analytis, J.-H. Chu, Z.K. Liu, S.-K. Mo, X.L. Qi, H.J. Zhang, D.H. Lu, X. Dai, Z. Fang, S.C. Zhang, I.R. Fisher, Z. Hussain, Z.-X. Shen, Experimental realization of a three-dimensional topological insulator, Bi2Te3. Science **325**(5937), 178–181 (2009)
21. D. Hsieh, D. Qian, L. Wray, Y. Xia, Y.S. Hor, R.J. Cava, M.Z. Hasan, A topological Dirac insulator in a quantum spin hall phase. Nature **452**(7190), 970–974, 04 (2008)
22. Y.S. Hor, A. Richardella, P. Roushan, Y. Xia, J.G. Checkelsky, A. Yazdani, M.Z. Hasan, N.P. Ong, R.J. Cava, p-type Bi2Se3 for topological insulator and low-temperature thermoelectric applications. Phys. Rev. B **79**, 195208 (2009)
23. D. Hsieh, Y. Xia, D. Qian, L. Wray, J.H. Dil, F. Meier, J. Osterwalder, L. Patthey, J.G. Checkelsky, N.P. Ong, A.V. Fedorov, H. Lin, A. Bansil, D. Grauer, Y.S. Hor, R.J. Cava, M.Z. Hasan, A tunable topological insulator in the spin helical Dirac transport regime. Nature **460**(7259), 1101–1105, 08 (2009)
24. S.R. Park, W.S. Jung, C. Kim, D.J. Song, C. Kim, S. Kimura, K.D. Lee, N. Hur, Quasiparticle scattering and the protected nature of the topological states in a parent topological insulator Bi_2Se_3. Phys. Rev. B **81**, 041405 (2010)
25. Y. Xia, D. Qian, D. Hsieh, R. Shankar, H. Lin, A. Bansil, A.V. Fedorov, D. Grauer, Y.S. Hor, R.J. Cava, M.Z. Hasan, Topological control: systematic control of topological insulator Dirac fermion density on the surface of Bi2Te3. arXiv, e-prints, July (2009)
26. D. Hsieh, Y. Xia, D. Qian, L. Wray, F. Meier, J.H. Dil, J. Osterwalder, L. Patthey, A.V. Fedorov, H. Lin, A. Bansil, D. Grauer, Y.S. Hor, R.J. Cava, M.Z. Hasan, Observation of time-reversal-protected single-Dirac-cone topological-insulator states in Bi_2Te_3 and Sb_2Te_3. Phys. Rev. Lett. **103**, 146401 (2009)
27. L. Fu, Topological crystalline insulators. Phys. Rev. Lett. **106**, 106802 (2011)
28. Y.J. Wang, W.-F. Tsai, H. Lin, S.-Y. Xu, M. Neupane, M.Z. Hasan, A. Bansil, Nontrivial spin texture of the coaxial Dirac cones on the surface of topological crystalline insulator SnTe. Phys. Rev. B **87**, 235317 (2013)

Chapter 2
Properties of Bi_2Se_3 and Bi_2Te_3

In this chapter I will present the fundamentals of the studied systems. I will begin by identifying the crystal structure of each system as well as their point and space group symmetries. From there I will move on to a review of measurements of the bulk vibrational structure using Raman, IR, and inelastic neutron scattering spectroscopies. Additionally, I will present recent ARPES measurements of the surface electronic structure that clearly indicate the presence of chiral DFQs, whose interaction with phonons is the main topic of this dissertation.

2.1 Crystal Structure

The strong 3D TIs Bi_2Se_3 and Bi_2Te_3 share the same rhombohedral structure, which is presented in Fig. 2.1. The bulk structure consists of alternating hexagonal monatomic crystal planes stacking in ABC order. Units of X-Bi-X-Bi-X (X = Se, Te) form quintuple layers (QLs): bonding between atomic planes within a QL is covalent whereas bonding between adjacent QLs is predominantly of the van der Waals type. This weak bonding between QLs allows the crystal to be easily cleaved along an inter-QL plane, a process which will be further elaborated upon in Chap. 4. The crystal structure belongs to the space group $R\bar{3}m$, while the point group contains a binary axis (with twofold rotation symmetry), a bisectrix axis (appearing in the reflection plane), and a trigonal axis (with threefold rotation symmetry).

It is convenient to work in the hexagonal basis when talking about this structure. A unit cell in the hexagonal basis contains 3 QLs and thus 15 atoms, whereas the

Fig. 2.1 Hexagonal unit cell of the Bi_2Te_3 crystal comprised of three QLs and belonging to the space group $R\bar{3}m$. Note that the Te2 layer within each QL is a center of inversion symmetry. Figure from [10]

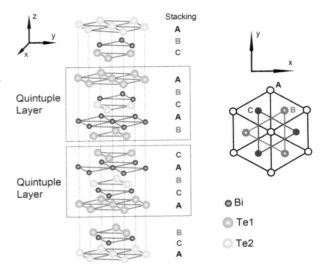

actual primitive cell in the rhombohedral basis contains five atoms. For now I will work in the hexagonal basis and write the translation vectors as

$$\mathbf{t}_1 = a\left(\frac{\sqrt{3}}{2}, -\frac{1}{2}, 0\right), \quad \mathbf{t}_2 = a(0, 1, 0), \quad \mathbf{t}_3 = c(0, 0, 1) \quad (2.1)$$

where a and c are lattice constants of the hexagonal cell. Using X as the subscript to indicate the material we have, $a_{Se} = 4.14\,\text{Å}$, $c_{Se} = 28.64\,\text{Å}$, $a_{Te} = 4.38\,\text{Å}$, and $c_{Te} = 30.49\,\text{Å}$. The corresponding reciprocal lattice vectors are

$$\mathbf{G}_1 = \frac{2\pi}{a}\left(\frac{2}{\sqrt{3}}, 0, 0\right), \quad \mathbf{G}_2 = \frac{2\pi}{a}\left(\frac{1}{\sqrt{3}}, 1, 0\right), \quad \mathbf{G}_3 = \frac{2\pi}{c}(0, 0, 1) \quad (2.2)$$

The reciprocal space structure of Bi_2X_3 is shown in Fig. 2.2. The first bulk BZ is actually presented for the rhombohedral basis and has an interesting shape with eight hexagonal faces and six rectangular faces. The surface reciprocal lattice, which is the primary focus of this dissertation, is a 2D hexagonal lattice obtained by taking the projection of the bulk BZ along the q_z axis as shown in Fig. 2.2. The SBZ contains three high-symmetry points including $\bar{\Gamma}$ at the zone center, $\bar{M}$ at the center of the zone edge, and $\bar{K}$ at the intersection of two zone edges. It is worth noting that not all $\bar{K}$ points are equivalent because they cannot be connected directly by a reciprocal lattice vector.

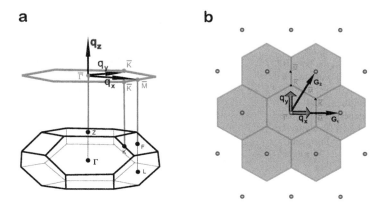

Fig. 2.2 Bulk and surface reciprocal space structure of Bi_2X_3. Panel (**a**) shows the rhombohedral bulk BZ and high-symmetry points. The SBZ is represented as a projection along $\mathbf{q}_z$. The extended surface reciprocal lattice is shown in (**b**) depicting high-symmetry points $\bar{\Gamma}, \bar{M}$, and $\bar{K}$ as well as reciprocal lattice vectors $\mathbf{G}_1$ and $\mathbf{G}_2$. Figure from [10]

2.2 Bulk Vibrational Structure

At this point I will present measurements of the bulk vibrational spectra of both Bi_2Se_3 and Bi_2Te_3. This data will allow me to fix the values of some of the empirical parameters entering the pseudocharge phonon model (to be discussed in Chap. 5) used to identify the character and symmetry of the bulk and surface phonon dispersions of the crystal. Investigations of the bulk vibrational structure have been carried out using one of the three methods: Fourier transform infrared (FTIR) spectroscopy, Raman spectroscopy, and inelastic neutron scattering spectroscopy. The first two methods listed are light scattering spectroscopies and thus are only capable of providing information about optical phonon modes at the BZ center because of the low momentum transfer involved in the scattering process. Neutron scattering, on the other hand, is capable of providing detailed information of bulk phonon dispersions off the Γ point because of neutron's significant momentum owing to its finite mass.

The primitive rhombohedral cell of Bi_2X_3 contains five distinct atoms, each with three degrees of freedom. Therefore we expect a total of fifteen phonon modes at any given wave vector, three acoustic and twelve optical. The atomic displacements for these modes at the Γ point are displayed in Fig. 2.3. The modes are labeled according to their symmetry with A modes belonging to a one-dimensional irreducible representation (irrep) and E modes belonging to a two-dimensional (doubly degenerate) irrep. The subscripts u and g indicate an even or odd parity about the central atom in the primitive cell. The former are accessible by FTIR scattering whereas the latter are only accessible via Raman scattering. Since the E modes are doubly degenerate the number of unique optical phonon frequencies is reduced from twelve to eight.

Fig. 2.3 Diagram depicting the atomic displacements of the five atoms in the primitive cell of Bi_2X_3 for optical phonon modes at the bulk BZ center. The modes are divided into distinct types: non-degenerate A_u and doubly degenerate E_u modes accessible by FTIR scattering and non-degenerate A_g and doubly degenerate E_g modes accessible by Raman scattering. Image taken from Ref. [1]

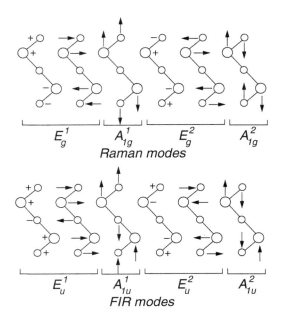

Table 2.1 Experimental values of the optical phonon frequencies of Bi_2Se_3 and Bi_2Te_3 at Γ. Data taken from [1]

Bi_2Se_3		Bi_2Te_3	
Mode	Frequency (cm^{-1})	Mode	Frequency (cm^{-1})
A^1_{1g}	72	A^1_{1g}	62.5
A^2_{1g}	174.5	A^2_{1g}	134
E^1_g	N/A	E^1_g	N/A
E^2_g	131.5	E^2_g	103
A^1_{1u}	N/A	A^1_{1u}	94
A^2_{1u}	N/A	A^2_{1u}	120
E^1_u	65	E^1_u	50
E^2_u	129	E^2_u	95

Studies of the bulk vibrational structure of Bi_2Se_3 are actually relatively few in number. To the best of my knowledge, no group has ever performed neutron scattering spectroscopy on these samples. Thus we are limited to results from studies using FTIR and Raman spectroscopy [1]. In their study, Richter and company were able to successfully observe only five of the eight unique optical modes at the Γ point. Their measured values can be found in Table 2.1.

In the case of the Bi_2Te_3 all three spectroscopies have been performed by various groups [1–4]. We thus have more complete description of the bulk vibrational

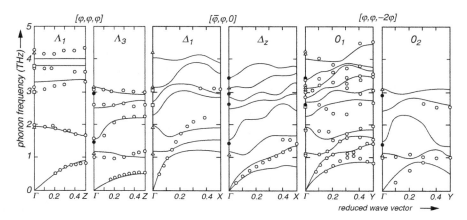

Fig. 2.4 Measurements of the bulk phonon dispersion of Bi_2Te_3. *Open circles* denote neutron data, *open squares* and *closed circles* correspond to FTIR data, and *open triangles* signify Raman data. The *solid lines* represent calculations based on a shell model for the phonon frequencies. Results are shown for the three distinct high-symmetry directions (Λ, Δ, and Σ) in the bulk BZ. Image taken from Ref. [2]

structure of this crystal. The data from FTIR and Raman spectroscopies can be found in Table 2.1. One will notice that all mode frequencies are smaller than their Bi_2Se_3 counterparts, consistent with heavier mass of Te when compared to Se. Neutron scattering data for this crystal is available in Fig. 2.4. Dispersions along three high-symmetry directions are shown, with two panels for each direction corresponding to distinct symmetry classes. This data will be fit in Chap. 5 to obtain values for the bulk force constant parameters appearing in our pseudocharge phonon model.

2.3 Electronic Structure

One of the most interesting aspects of the topological insulators Bi_2Se_3 and Bi_2Te_3 is their unique electronic structure. Both materials exhibit semiconducting behavior, with bulk band gaps of approximately 300 meV in the case of Bi_2Se_3 and 100–150 meV for Bi_2Te_3. Both possess an inverted band structure owing to the presence of strong spin-orbit coupling resulting from the large Bi mass. This, along with the inversion symmetry of the bulk crystal, guarantees a non-trivial $\mathbb{Z}_2$ as mentioned in Chap. 1 making both strong 3D topological insulators. The surfaces of the crystal are thus host to metallic DFQ surface states protected by TRI which serve to close the bulk band gap and usher in the change in $\mathbb{Z}_2$ occurring at the crystal/vacuum interface.

Because electronic transport experiments have difficulty distinguishing surface state contributions to the conductivity from imperfections of the bulk crystal (including Se and Te vacancies), experimental verification of said surface states has largely been carried out using ARPES. Numerous studies of the surface band structure have confirmed the existence of a single, linearly dispersive Dirac cone

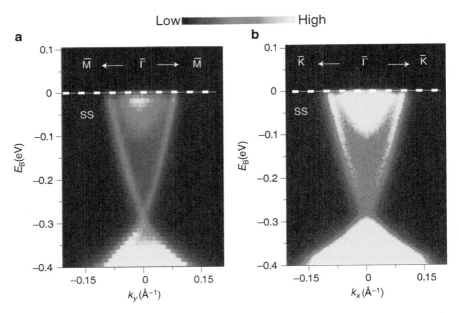

Fig. 2.5 ARPES data illustrating the presence of a single, linearly dispersive Dirac cone on the surface of Bi_2Se_3 centered at the $\bar{\Gamma}$ point. The surface states appear as two branches emanating from the Dirac point at -300 meV and continuing upward to the Fermi level. Contributions from the filled bulk valence band and partially filled bulk conduction appear as bright patches of *yellow* over a range of crystal momentum. Image taken from Ref. [5]

centered about the $\bar{\Gamma}$ point in the SBZ for both materials. Example ARPES data for Bi_2Se_3 can be seen in Fig. 2.5. Whereas the Dirac cone is isotropic in Bi_2Se_3, the Fermi velocity modulates slightly depending on the crystal momentum direction in the case of Bi_2Te_3. This gives rise to a warping effect which gives the Fermi surface a star-like shape at energies high relative to the Dirac point in contrast to Bi_2Se_3 where the Fermi surface remains nearly circular for all energies within the bulk band gap.

In both materials the metallic surface states are also spin-polarized [6, 7], again owing to the significant spin–orbit interaction. Specifically, the spin of the electronic state is always locked perpendicular to the crystal momentum and lies in the surface plane (see Fig. 2.6). Thus, these materials have a definite chirality to their surface electronic states because of the spin modulation that occurs when traversing the curve in 2D reciprocal space defined by the Fermi surface. Interestingly enough, this leads to a non-trivial Berry's phase [8, 9] of π when an electronic state is taken about the Fermi surface, which is related to their topological character. This is quite different from a non-magnetic metal wherein each state at any given crystal momentum possesses a two-fold spin degeneracy. Hence, in some ways the surface of a topological insulator can be viewed as half of a non-magnetic metal, with only a single spin species occurring at a given crystal momentum.

2.3 Electronic Structure

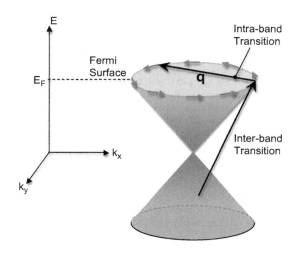

Fig. 2.6 Schematic diagram of the idealized Dirac cone electronic structure present on the surfaces of Bi_2Se_3 and Bi_2Te_3. The spin texture of the Fermi surface is shown by the *red arrows*. A hypothetical low-energy scattering event involving momentum transfer **q** (perhaps from a phonon) is shown. Inter-band transitions involving higher energy are also possible. Figure from [10]

The spin-momentum locking in the surface states of topological insulators has interesting consequences. First and foremost, states on opposite sides of the Fermi surface which are propagating in opposite directions (owing to the reversed sign of the group velocity upon the substitution $\mathbf{k} \to -\mathbf{k}$) cannot backscatter into each other in the absence of magnetic perturbations. This is because such a scattering event would necessarily require flipping the spin. More generally speaking, backscattering of the electronic states on the surfaces of these materials is suppressed in the presence of time-reversal invariant perturbations. Without backscattering, the DFQs on the surface are immune to localization, a trait that has profound implications for electronic transport and potential device applications. Second, because states with opposite group velocity have opposite spin but the same charge, one can theoretically produce a net spin current without a net movement of charge by simply populating states with opposite crystal momentum simultaneously. This is quite similar to what happens in the QSHE except that here the transport takes place in a 2D plane where the Fermi surface contains a continuum of electronic states rather than a 1D edge where the Fermi surface is simply two points in reciprocal space. For these reasons, TIs are garnering interest as potential materials to be used in emerging spintronic technologies. However, to truly be effective, scattering of DFQs from time reversal invariant perturbations (phonons, crystal vacancies, grain boundaries, etc.) should also be small. Otherwise, although exact localization may not occur, the favorable transport characteristics on a macroscopic scale could be adversely affected. As such, the present study sets out to determine the nature and degree of DFQ-phonon coupling on the surfaces of TIs using helium atom surface scattering spectroscopy.

References

1. W. Richter, C.R. Becker, A raman and far-infrared investigation of phonons in the rhombohedral $V_2 - VI_3$ compounds Bi_2Te_3, Bi_2Se_3, Sb_2Te_3 and $Bi_2(Te_{1-x}Se_x)_3$ ($0 < x < 1$), $(Bi_{1-y}Sb_y)_2Te_3$ ($0 < y < 1$). Phys. Status Solidi (B) **84**(2), 619–628 (1977)
2. W. Kullmann, G. Eichhorn, H. Rauh, R. Geick, G. Eckold, U. Steigenberger, Lattice dynamics and phonon dispersion in the narrow gap semiconductor Bi2Te3 with sandwich structure. Phys. Status Solidi (B) **162**(1), 125–140 (1990)
3. W. Kullmann, J. Geurts, W. Richter, N. Lehner, H. Rauh, U. Steigenberger, G. Eichhorn, R. Geick, Effect of hydrostatic and uniaxial pressure on structural properties and Raman active lattice vibrations in Bi2Te3. Phys. Status Solidi (B) **125**(1), 131–138 (1984)
4. V. Wagner, G. Dolling, B.M. Powell, G. Landweher, Lattice vibrations of Bi2Te3. Phys. Status Solidi (B) **85**(1), 311–317 (1978)
5. Y. Xia, D. Qian, D. Hsieh, L. Wray, A. Pal, H. Lin, A. Bansil, D. Grauer, Y.S. Hor, R.J. Cava, M.Z. Hasan, Observation of a large-gap topological-insulator class with a single dirac cone on the surface. Nat. Phys. **5**(6), 398–402, 06 (2009)
6. D. Hsieh, Y. Xia, D. Qian, L. Wray, J.H. Dil, F. Meier, J. Osterwalder, L. Patthey, J.G. Checkelsky, N.P. Ong, A.V. Fedorov, H. Lin, A. Bansil, D. Grauer, Y.S. Hor, R.J. Cava, M.Z. Hasan, A tunable topological insulator in the spin helical dirac transport regime. Nature **460**(7259), 1101–1105, 08 (2009)
7. Z. Xie, S. He, C. Chen, Y. Feng, H. Yi, A. Liang, L. Zhao, D. Mou, J. He, Y. Peng, X. Liu, Y. Liu, G. Liu, X. Dong, L. Yu, J. Zhang, S. Zhang, Z. Wang, F. Zhang, F. Yang, Q. Peng, X. Wang, C. Chen, Z. Xu, X.J. Zhou, Orbital-selective spin texture and its manipulation in a topological insulator. Nat. Commun. **5**, 02 (2014)
8. M.Z. Hasan, C.L. Kane, Colloquium. Rev. Mod. Phys. **82**, 3045–3067 (2010)
9. Y. Ando. Topological insulator materials. J. Phys. Soc. Jpn. **82**(10), 102001 (2013)
10. C. Howard, M. El-Batanouny, R. Sankar, F.C. Chou, Anomalous behavior in the phonon dispersion of the (001) surface of Bi2Te3 determined from helium atom-surface scattering measurements. Phys. Rev. B **88**, 035402 (2013)

Chapter 3
Helium Atom-Surface Scattering (HASS)

This chapter will focus on the HASS technique and its advantages over other surface probes. From there I will move on to describing the manner in which the atom–surface interaction is modeled. Lastly, I will describe the scattering processes that can occur and what information they carry about the structure and vibrational character of the surface.

3.1 The Benefits of HASS

Helium atom-surface scattering is a high-resolution technique capable of providing valuable information about surface structure and dynamics of metallic and insulating compounds [1–4]. The technique relies on the production of a monoenergetic helium beam probe that is scattered from solid surfaces, either elastically or inelastically. HASS has several advantages over other surface science techniques such as EELS:

1. The helium atoms employed are at thermal energies, making them non-destructive.
2. Helium's closed K shell makes it chemically inert, decreasing the risk of chemisorption when it interacts with the surface.
3. The scattering of thermal helium atoms actually results from the overlap between the atomic electron orbitals with the surface electronic charge density. Thus the classical turning point for the incident thermal beam actually lies 2–3 Å above the terminal ion layer of the solid [2], implying no penetration of the probe into the material. Hence HASS is exclusively surface sensitive.
4. The characteristic de Broglie wavelength of thermal helium atoms lies in the range $0.5 < \lambda < 1.5$ Å, yielding wave-numbers comparable to typical SBZ dimensions.

5. High intensity beams are easily produced with very high energy resolution (≤ 1 meV).
6. The energy of thermal helium atoms ranges from 10 to 80 meV, well-matched to the excitation energies of surface phonons. Moreover, because the energies are low, multi-phonon events are suppressed.
7. Since helium is the smallest atom with a diameter of roughly 0.5 Å, scattering from a solid usually only involves a single ion core on the surface. This reduces complications arising from multi-scattering processes.

3.2 The Surface Interaction Potential

Applying neutral particle probes to study surface structure and dynamics requires the knowledge of the particle–surface interaction on the microscopic scale. I begin by considering a solid with semi-infinite slab geometry occupying the region $z < 0$. The helium–surface interaction potential consists of two parts, one attractive and one repulsive, originating from distinct physical phenomena.

Far from the surface, van der Waals forces dominate causing helium atoms to experience a mild attractive potential [5] proportional to r^{-6}. Using cylindrical coordinates and integrating this interaction potential over the entire crystal slab for a helium atom at a height z above the crystal surface we find

$$V_{\text{att}}(\mathbf{r}) = -\int_{z}^{\infty} dz' \int_{0}^{\infty} d\rho \int_{0}^{2\pi} d\phi \, \frac{C}{(\rho^2 + z'^2)^3} \sim \frac{-C}{z^3} \quad (3.1)$$

where the constant C depends on the surface in question. Thus, far from the surface the helium atoms experience a cubic attractive potential.

However, close to the surface, overlap between the atomic electron orbitals and those of the surface charge density create an abrupt repulsive potential. The wave functions of the incoming helium atom's electrons tend to orthogonalize with those of the surface electron charge density $\rho(\mathbf{r})$. This creates an increase in kinetic energy of the combined system that leads to a repulsive potential. A previous study [6] showed that, to good approximation, this repulsive potential is linear in the surface charge density

$$V_{\text{rep}}(\mathbf{r}) = \alpha \rho(\mathbf{r}) \quad (3.2)$$

Here α is a constant typically on the order of 10^2 eVa_0^3 where a_0 is the Bohr radius.

The attractive and repulsive components of the atom–surface interaction combine to yield a curve with a minimum in the region $z = 5 - 10$ Å depending on the exact chemical nature of the surface in question, as can be seen in Fig. 3.1. Strictly speaking, incident helium atoms can become trapped in the bound states formed by this potential well. This can happen by one of two ways. First, the periodicity of the crystal parallel to the surface can reorient the helium momentum such that it is

3.2 The Surface Interaction Potential

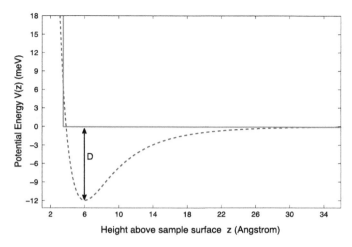

Fig. 3.1 Schematic of the helium–surface interaction potential experienced by the incident helium beam. The *blue curve* qualitatively traces out the true interaction potential. The *red curve* is an approximation known as the hard corrugated wall model, which is valid when the incident helium energy is much larger than the well depth D

directed parallel, rather than perpendicular, to the surface. This can create a scenario in which an elastically scattered helium atom propagates along the periodic potential of the surface, much like an evanescent wave. Second, an incident helium atom may lose some energy upon interacting with the surface (for example, by creating a phonon), and become trapped by the attractive potential.

The trapping probability described in both of these scenarios is reduced greatly if the helium beam employed has sufficiently high initial kinetic energy and sufficiently low incident angle (as measured from the surface normal). In fact, if the energy is high relative to the well depth D, one can ignore the attractive part of the potential entirely and instead work within the *hard corrugated wall model*. In this approximation, the interaction potential has the form

$$V(z) = \begin{cases} 0, & z > \zeta(\mathbf{R}) \\ \infty, & z \leq \zeta(\mathbf{R}) \end{cases} \quad (3.3)$$

Here, $\zeta(\mathbf{R})$ is known as the corrugation function. One can think of it as an effective surface height experienced by the helium atom at the position $\mathbf{R} = (x, y)$. In this approximation then, incident helium atoms travel as free particles until they reach $z = \zeta(\mathbf{R})$ at which point they scatter, either elastically or inelastically. Recalling our assumption that the repulsive potential is linear in the surface charge density, one can see that $\zeta(\mathbf{R})$ is equivalent to a constant surface charge density contour as depicted in Fig. 3.2. The larger the component of helium momentum normal to the sample surface, the deeper the atom will into these contours, thus sampling a more corrugated surface. Since this corrugation reflects the arrangement of the underlying ion cores, HASS can directly probe the atomic arrangement and topology of the sample surface.

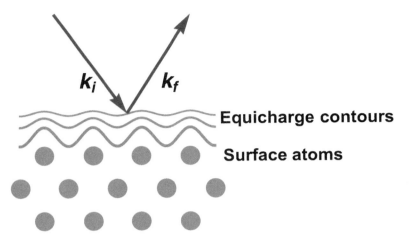

Fig. 3.2 Cross section of a 3D crystal lattice. Depicted is a process in which a helium atom with incident wave vector k_i scatters from a constant charge density contour into a state of wave vector k_F

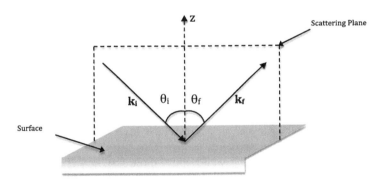

Fig. 3.3 Diagram of the in-plane scattering geometry. Note that θ_i need not necessarily equal θ_f for either elastic or inelastic scattering processes described in the text

3.3 The Kinematics of HASS

3.3.1 Elastic Scattering

To begin discussing the helium scattering process quantitatively I consider the in-plane scattering geometry shown in Fig. 3.3. In the hard corrugated wall model both the incident and scattered helium beams may be treated as free particles, allowing one to write

3.3 The Kinematics of HASS

$$E_i = \frac{\hbar k_i^2}{2m} \qquad E_f = \frac{\hbar k_f^2}{2m} \qquad (3.4)$$

In the case of elastic scattering the initial and final energies are the same or, equivalently, the incident and scattered wavevectors have the same magnitude. However, due to the periodic invariance parallel to the surface, the component of the scattered wavevector parallel to the surface need only be conserved up to a reciprocal lattice vector of the surface lattice. This allows us to write

$$k_f \sin \theta_f = k_i \sin \theta_f = k_i \sin \theta_i \pm n G_0 \qquad (3.5)$$

where I used the fact that $k_f = k_i$. G_0 is a primitive reciprocal lattice vector along a high-symmetry direction in the SBZ, and n can take on any integer value. Solving Eq. (3.5) for θ_f we find

$$\sin \theta_f = \sin \theta_i \pm n \frac{G_0}{k_i} \qquad (3.6)$$

which defines the locations of the elastic diffraction peaks.

In practice, elastic diffraction studies performed using HASS can serve different purposes. The most basic use is that of a diagnostic tool for surface orientation; peak locations in a diffraction pattern can be used as input to Eq. (3.6), which is then solved for the reciprocal lattice vector. Knowledge of G_0 determines the relative orientation of the sample surface and scattering plane, which is essential when measuring phonons along particular high-symmetry directions as described below. Additionally, although it is not the main topic of this dissertation, thorough analysis of the diffraction peak intensities along a particular high-symmetry direction can allow one to reconstruct the surface corrugation $\zeta(\mathbf{R})$ using Fourier analysis. This allows one to perform detailed measurements of real space surface topography.

3.3.2 Inelastic Scattering and Time-of-Flight Technique

In the case of inelastic scattering the energies of the incident and scattered helium beams will differ. This cannot occur in the case of a rigid surface corrugation. However, if one admits the possibility of a deformable $\zeta(\mathbf{R}, t)$ the incident helium beam can transfer some of its energy into vibrational energy associated with the deformation of the surface charge density contours. If this charge density deformation couples to the motion of the underlying ions, the ionic positions can be disturbed from equilibrium and a phonon is created. The inverse of this process can also occur wherein energy from the oscillating surface charge density is transmitted to the incident helium beam, which then leaves with a larger kinetic energy upon scattering from the surface.

Again assuming that both the incident and scattered helium beams behave as free particles, the change in energy of the beam is

$$\Delta E = \frac{\hbar^2}{2m}(k_f^2 - k_i^2) = E_i\left(\frac{k_f^2}{k_i^2} - 1\right) \tag{3.7}$$

The wavevector of the scattered helium will have a different magnitude than that of the incident beam. However, we can write without ambiguity that the change in the component of the wavevector parallel to the surface plane ΔK is

$$\Delta K = k_f \sin\theta_f - k_i \sin\theta_i \tag{3.8}$$

Rearranging for k_f we find

$$k_f = \frac{\Delta K + k_i \sin\theta_i}{\sin\theta_f} \tag{3.9}$$

We can plug Eq. (3.9) into Eq. (3.7) and rearrange to find

$$\Delta E = E_i\left(\frac{\sin^2\theta_i}{\sin^2\theta_f}\left(\frac{\Delta K}{k_i \sin\theta_i} + 1\right)^2 - 1\right) \tag{3.10}$$

Thus we find that the change in beam energy is parabolic in the change in the parallel component of the momentum. This change in beam energy may be positive or negative, corresponding to the annihilation or creation of a surface phonon, as described at the opening of this subsection. The same conservation rule for ΔK that we found for the case of elastic scattering still holds except we must now account for the momentum of the phonon. Thus we have

$$\Delta K = q \pm nG_0 \tag{3.11}$$

where q is the phonon wavevector. Lastly, conservation of energy clearly implies that

$$\Delta E = \pm \hbar\omega(\mathbf{q}) \tag{3.12}$$

which merely says that the positive or negative changes in beam energy (corresponding to phonon annihilation and creation, respectively) must come in increments of the phonon energy.

Equation (3.10) gives us the so-called *scan curves*. They represent the kinematically allowed changes in beam energy when momentum conservation is accounted for. One will notice that, for fixed E_i and θ_i, the scan curves form a family of parabolas, each with a distinct value of θ_f. We can imagine superimposing these

3.3 The Kinematics of HASS

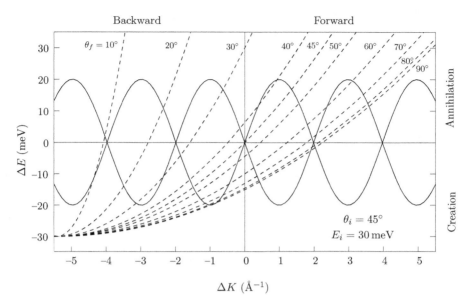

Fig. 3.4 Overlay of a family of scan curves (*dashed lines*) for fixed E_i and θ_i atop an example dispersion relation (*solid curves*). Notice that the dispersion must also be reflected about the ΔK axis to account for phonon creation events (i.e., $\Delta E < 0$)

parabolas over the phonon dispersion relation of a crystal along a particular high-symmetry direction in the SBZ. An example of this is provided in Fig. 3.4 where the sample dispersion is that of a 1D monoatomic chain of ions. The intersections of the dispersion curves (which are the desired unknown in the experiment) and scan curves correspond to measurable scattering events.

Determining the surface phonon dispersion requires the knowledge of a set of points of the form $(\mathbf{q}, \hbar\omega(\mathbf{q}))$. However, one should notice that it is sufficient to measure ΔE of the helium beam alone. This information, in conjunction with θ_i, θ_f, E_i, and G_0 (obtained from preliminary elastic diffraction measurements) is sufficient to determine $\mathbf{q}$ and $\hbar\omega(\mathbf{q})$ using Eqs. (3.10)–(3.12). In practice, measurement of the change in beam energy is accomplished by *time-of-flight* techniques whereby the scattered helium beam is allowed to travel over a fixed distance l and the arrival time at the detector is measured using a timing mechanism. Treating the helium atoms as classical particles one can write

$$\Delta E = \frac{m}{2}\left(\left(\frac{l}{t_f}\right)^2 - \left(\frac{l}{t_0}\right)^2\right) = \frac{m}{2}\left(\frac{l}{t_0}\right)^2\left(\left(\frac{t_0}{t_f}\right)^2 - 1\right) = E_i\left(\left(\frac{t_0}{t_f}\right)^2 - 1\right) \tag{3.13}$$

where m is the mass of the helium atom, t_0 is the arrival time of the elastically scattered helium, and t_f is the arrival time of the inelastically scattered helium. Therefore, by measuring t_f for a particular inelastic scattering event, one obtains a single point $(\mathbf{q}, \hbar\omega(\mathbf{q}))$ in the surface phonon dispersion. In practice, a typical HASS experiment involves fixing E_i and θ_i and taking time-of-flight data at varying θ_f by moving the detector around the sample until a sufficient density of points is accrued to infer the actual dispersion curves of surface phonons along a high-symmetry direction of the SBZ.

References

1. J.P. Toennies, Helium atom scattering: a gentle and uniquely sensitive probe of surface structure and dynamics. J. Phys. Condens. Matter **5**(33A), A25 (1993)
2. D. Farias, K.-H. Rieder, Atomic beam diffraction from solid surfaces. Rep. Prog. Phys. **61**(12), 1575 (1998)
3. J. Braun, P. Ruggerone, G. Zhang, J.P. Toennies, G. Benedek, Surface phonon dispersion curves of thin Pb films on Cu(111). Phys. Rev. B **79**, 205423 (2009)
4. A. Tamtögl, P. Kraus, M. Mayrhofer-Reinhartshuber, D. Campi, M. Bernasconi, G. Benedek, W.E. Ernst, Surface and subsurface phonons of Bi(111) measured with helium atom scattering. Phys. Rev. B **87**, 035410 (2013)
5. E. Zaremba, W. Kohn, Van der waals interaction between an atom and a solid surface. Phys. Rev. B **13**, 2270–2285 (1976)
6. N. Esbjerg, J.K. Nørskov, Dependence of the he-scattering potential at surfaces on the surface-electron-density profile. Phys. Rev. Lett. **45**, 807–810 (1980)

Chapter 4
Experimental Apparatus and Technique

4.1 Surface Laboratory Facilities

The main experimental apparatus of the Laboratory for Surface Physics and Electron Spectroscopies at Boston University consists of a series of HV and UHV chambers equipped with HASS capabilities and diagnostic tools including a LEED unit. An aerial schematic of the entire setup can be seen in Fig. 4.1. In the proceeding sections I will detail the different parts of the apparatus and their function. From there, I will present the sample preparation methodology and measurement techniques for both inelastic and elastic scattering.

4.2 Source Chamber

Successful implementation of the HASS technique requires the generation of a monoenergetic helium beam via a continuum jet expansion monochromator visible on the left-hand side of Fig. 4.1. The source chamber is maintained at a background pressure P_b via a 12,500 L/s diffusion pump backed by a roots pump, which is in turn backed by a mechanical pump. An external tank supplies high pressure ($P_0 \sim 500$ psi), high purity (99.999 %) helium to the monochromator assembly within the source chamber. The monochromator itself consists of a temperature controlled helium reservoir with a nozzle on one end that has a small orifice of diameter $d = 20\,\mu$m. The helium supplied to the reservoir by the external tank thermalizes with the reservoir walls at temperature T_0 and then undergoes adiabatic expansion as it escapes into the source chamber through the orifice. A depiction of this process can be seen in Fig. 4.2.

The adiabatic expansion causes rapid cooling of the helium gas, typically on the order of 10^9 K/s leading to a new beam temperature T. This cooling leads to a drastic

24   4 Experimental Apparatus and Technique

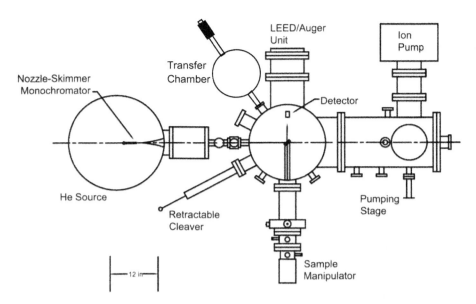

Fig. 4.1 Aerial view of the HASS facility at Boston University

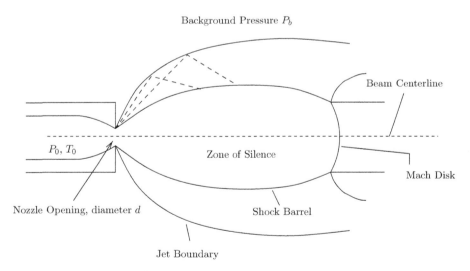

Fig. 4.2 Schematic illustrating the adiabatic expansion process of the monochromator. Helium within the zone of silence has a very narrow speed (and hence energy) distribution and is extracted for experimental use by a skimmer assembly (not shown)

4.2 Source Chamber

collapse in the variance of the Maxwell speed distribution of the gas atoms, creating the monoenergetic beam desired for the experiment. The terminal velocity of the beam after leaving the nozzle is

$$v = \sqrt{\frac{5k_B T_0}{m}} \qquad (4.1)$$

One should notice that the numerical coefficient differs from simple equipartition arguments due to the fact that the helium is undergoing a dynamic process and is not in equilibrium. The expansion reorients the random thermal velocities of the helium within the reservoir into a direction perpendicular to plane of the nozzle orifice. One can further characterize the beam by examining the variance of the component of the velocity parallel to the beam direction, given by Martini [1]

$$\delta v = 2\sqrt{\frac{(2\ln 2)k_B T}{m}} \qquad (4.2)$$

Finally, the speed ratio[2] may be defined as

$$S = \sqrt{\frac{mv^2}{2k_B T}} = \sqrt{\frac{5T_0}{2T}} \qquad (4.3)$$

The speed ratio can be used to easily quantify the resolution of the outgoing beam via

$$\frac{\delta v}{v} = \frac{2\sqrt{\ln 2}}{S} = \frac{1.67}{S} \qquad (4.4)$$

Thus a high resolution beam is created by operating at the largest S possible[2], which is in turn controlled by the product $P_0 d$. Theoretical predictions of the speed ratio from both classical and quantum mechanical models are shown in Fig. 4.3. Although the classical model predicts a linear dependence of the speed ratio on the product $P_0 d$ one can see that the quantum mechanical model exhibits a significant deviation from linearity. This deviation is the result of a phenomena known as *zero energy resonance* resulting from the fact that the He dimer has a very low (10^{-7} eV) bound state energy. This low energy bound state increases the scattering cross section of the beam appreciably and enhances the speed ratio, helping improve the velocity resolution of the beam even further. In practice the speed ratio (and hence velocity resolution) is limited by the nozzle diameter d and P_0, which cannot be made arbitrarily high without exceeding the pumping capacity of the diffusion pump and roots pumps attached to the source chamber. Typical values are $\delta v/v \approx 1\%$.

The rapid cooling of the helium gas has other interesting hydrodynamic implications. Indeed, the sound velocity of the helium gas plume emitted from the nozzle, being proportional to $\sqrt{T}$, is also drastically reduced by the expansion process. The resulting scenario is then a plume of helium gas whose constituent atoms are

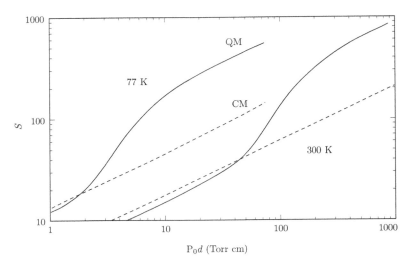

Fig. 4.3 Dependence of the speed ratio on $P_0 d$ for liquid nitrogen and room temperatures. Classical and quantum models are given by the *dashed* and *solid lines*, respectively. Figure from [2]

supersonic, traveling faster than the sound velocity of the medium itself. As such, a shock barrel or sound barrier forms towards the edges of the plume as depicted in Fig. 4.2. Helium atoms that reach the shock barrel will experience a temperature increase and return to a diffusive flow regime rather than the hydrodynamic flow characterizing the zone of silence, making them unsuitable for the experiment. However, the supersonic helium in the zone of silence maintains its monoenergetic character and is allowed to flow throw a skimmer assembly before reaching the shock barrel. From here it passes into a buffer HV chamber where the flow transforms from hydrodynamic to molecular.

4.3 Target Chamber

After passing through the skimmer assembly the helium beam enters the target chamber of the apparatus. This is where the sample under study is housed and the actual scattering of the incident helium beam takes place. Due to novel detection techniques to be described later, this chamber also houses the detector.

4.3.1 Production and Monitoring of UHV

It has been well established[3] that HASS measurements are extremely sensitive to contamination of the sample surface by adsorbed residual gases in the UHV environment. For this reason it is imperative to keep the sample surface as free of adsorbates as possible, which is accomplished by reducing the pressure in the target chamber by a variety of pumping mechanisms. The target chamber is maintained at a base pressure of approximately 10^{-8} Torr via a turbomolecular pump backed by a mechanical pump.

Once this combination reaches its minimum attainable pressure a cryo-pumping routine begins. Liquid nitrogen is pumped into a hollow baffle occupying the inner wall of the pumping stage chamber depicted in Fig. 4.1. The walls of the baffle thermalize with the liquid nitrogen and start to act as a cold surface onto which residual gases adsorb, thus removing them from the UHV environment. This is particularly effective at reducing the partial pressure of water, one of the main contaminants in the chamber, into the 10^{-11} Torr range.

Finally, a titanium pump is used as a final measure to reduce the chamber pressure. A titanium filament, also located in the pumping stage chamber, is supplied with a strong AC current causing an increase in temperature that causes some of the filament to vaporize. Titanium settles on the cold nitrogen baffle surface and acts as a getter, increasing the probability that gases incident on the baffle walls will stick. The combination of these different pumping stages effectively reduces the partial pressures of the main contaminants in the chamber (H_2, H_2O, O_2, CO_2), as recorded on the attached Stanford Research Systems Residual Gas Analyzer, to 5×10^{-11} Torr. This allows one to maintain an initially clean sample surface for roughly 8–12 h during which HASS measurements can be performed.

4.3.2 Sample Manipulator

After entering the target chamber the first thing the helium beam encounters is the sample surface. Samples prepared by external collaborators[1] are cut into wafers and mounted on an OFHC copper post via UHV compatible conductive epoxy. A small aluminum cleaving pin is then attached to the sample surface with the same epoxy. This sample-post assembly is then fed into the target chamber via a retractable transfer arm located in the transfer chamber shown in Fig. 4.1. The sample-post assembly locks into the sample manipulator atop the sample stage via support fins, whereupon the transfer arm is removed.

The sample manipulator allows full control of the sample orientation via five separate degrees of freedom. There are three linear controls allowing for x,y,z motion of the sample, which are useful for the cleaving process and beam approach.

---

[1]Samples were prepared by Dr. F.C. Chou and Dr. R. Sankar of the Center of Condensed Matter Sciences at the National Taiwan University.

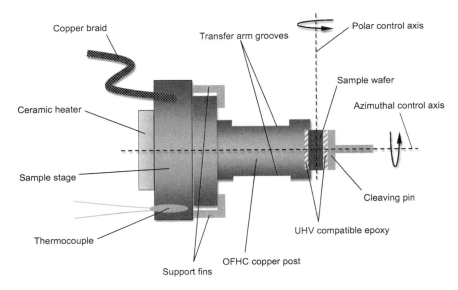

Fig. 4.4 Diagram of the sample-post assembly sitting atop the sample stage after being loaded into the target chamber via the transfer arm

Polar angle control allows one to adjust the angle θ_i depicted in Fig. 3.3 between the sample surface normal and the incident helium beam. Finally, azimuthal control allows the sample surface to rotate in a plane perpendicular to the surface normal, and assists in aligning the incident beam along a surface high-symmetry direction of the crystal.

The manipulator is also equipped with temperature control. On one side of the sample stage is a copper braid that is connected to an APD closed cycle helium refrigerator, which provides cooling capabilities. Additionally, the rear of the sample stage has a ceramic heater attached. Finally, a thermocouple attached to the side of the stage opposite to the copper braid allows for measurements of the sample temperature and acts as a feedback sensor for the external temperature controller unit. A detailed schematic of the sample-post assembly can be found in Fig. 4.4

4.3.3 Sample Cleaver

In order to ensure an initially clean sample surface for conducting the HASS measurements the samples must be cleaved in-situ. This is accomplished by a retractable cleaver, also pictured in Fig. 4.1. Once the sample has been secured to the sample manipulator, the cleaving arm can be moved toward the sample surface. When the cleaver is in close proximity to the cleaving pin, an abrupt strike with the spring-loaded blade knocks the cleaving pin off the sample surface, peeling away a piece of the sample and exposing a fresh surface. Due to the QL structure of Bi_2Te_3 and Bi_2Se_3 (see Fig. 2.1), the peeling process occurs quite naturally along the weak

4.3 Target Chamber

inter-QL bonds, ensuring that Te/Se is the terminal surface layer. For harder ionic crystals the cleaving pin methodology is often unsatisfactory because the crystals do not respond well to the combination of tensile and sheer stress in the peeling process. Instead, the crystal wafer itself must be hit with the cleaving blade along a particular crystal plane, causing it to crack and expose a fresh surface.

4.3.4 Helium Detector

4.3.4.1 Continuous (Elastic) Detection

Once the helium beam is scattered from the surface of the sample it must be recorded. This is accomplished via a two-stage process involving an electron beam and a multi-channel plate electron multiplier (MCPEM). First, the scattered helium beam passes through an angle-resolving orifice in an electron gun mounted to the detector carriage. This gun produces a collimated beam of monoenergetic electrons via thermionic emission from a low work function matrix cathode that travel perpendicular to the direction of motion of the scattered helium. The cathode voltage and electron optics within the gun are tuned to give the electrons just the right energy to excite a helium atom into the metastable $^3S_1$ excited state upon collision. The excited helium atoms exit the electron gun and collide with the MCPEM. An electron from the plate fills the vacant 1S energy level of the metastable helium atom imparting its excess energy to the 2S shell electron which is ejected it in a manner similar to Auger emission. This released electron propagates and multiplies through the channel plate resulting in an amplified pulse signal, which is ultimately read by the electronics of the detector. A schematic of the entire process can be found in Fig. 4.5.

4.3.4.2 Time-of-Flight (Inelastic) Detection

The detection scheme for the inelastic helium scattering is somewhat different than that of the elastic scattering. In this case the flight time of a given helium atom, from its excitation in the electron gun to its collision with the MCPEM, must be recorded in order to calculate its change in kinetic energy as described in Sect. 3.3.2. In practice this is done by creating a gate function in time in the electron gun that produces packets of scattered excited helium atoms, which are then allowed to traverse the fixed distance between the gun and MCPEM. A beam packet contains many atoms having different velocities corresponding to inelastic as well as elastic scattering events at the sample surface. The atoms disperse as they travel toward the MCPEM. Atoms reaching the detector early have gained energy due to phonon annihilation events at the sample surface, whereas the slow ones were involved in phonon creation events. In order to resolve the differences in helium atom arrival times, the opening time of the gate function should be roughly 1 % of the total time it

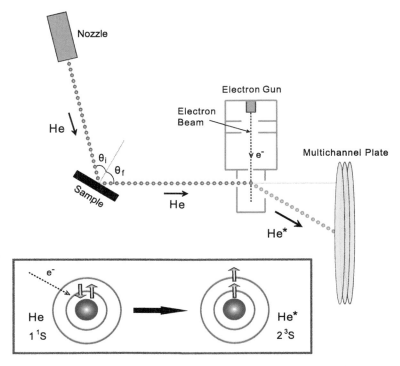

Fig. 4.5 Diagram of the detection scheme. The scattered helium beam is subjected to a perpendicular electron beam which excites it into a metastable state, allowing it to be detected by the multi-channel plate. Figure from [5]

takes the packet to arrive at the MCPEM. For a traditional mechanical beam chopper consisting of a notched, rotating wheel, minimum opening times are on the order or 1ms, ultimately constrained by the rotation speed which cannot exceed structural limitations of the material. At thermal helium velocities on the order of 10^3 m/s, a path length of 2–10 m is therefore needed to resolve arrival times. However, at the Boston University Laboratory for Surface Physics and Electron Spectroscopies, the gating is performed with the electron gun itself, which can be pulsed on much finer time scales than any mechanical chopper. The gate function opening time is reduced to the order of 1 μs, reducing the required flight path to roughly 7 cm.

The process outlined above is complicated by two factors. First, one may object that the collision of the helium beam with the electron beam will alter its momentum and thus affect the timing measurement. However, the orientation of the electron beam guarantees that only the component of the helium's momentum perpendicular to its original line of motion is affected. The original component of the helium momentum is preserved and thus the timing measurement is unaffected. Second, and perhaps more serious, is the fact that consecutive packets can have spatial (and hence temporal) overlap upon arriving at the MCPEM if the gate openings are close together in time. The easiest remedy would be to space out the gate opening times so that, at thermal velocities, there is no risk of overlap for the given flight path. However, this is not possible because the data collection times necessary to acquire a detectable number of inelastic events at any given θ_f would become prohibitively long, running into the limitations imposed by adsorbate exposure times described earlier. The ingenious solution is actually to use a pseudo-random gate function [4]. This allows one to deconvolute the overlapping packets and thus retain a high fidelity timing measurement. This pseudo-random gate function is comprised of a set of pulses whose total integrated opening time is roughly 10–50 % of the helium arrival time.

References

1. K. Martini, Elastic and inelastic helium beam scattering from Cu(1 0 0) surfaces and vibrational modes of incommensurate systems. Ph.D. thesis, Boston University (1986)
2. J.P. Toennies, K. Winkelmann, Theoretical studies of highly expanded free jets: influence of quantum effects and a realistic intermolecular potential. J. Chem. Phys. **66**(9), 3965–3979 (1977)
3. D. Farias, K.-H. Rieder, Atomic beam diffraction from solid surfaces. Rep. Prog. Phys. **61**(12), 1575 (1998)
4. G. Comsa, R. David, B.J. Schumacher, Magnetically suspended cross-correlation chopper in molecular beam-surface experiments. Rev. Sci. Instrum. **52**(6), 789–796 (1981)
5. C. Howard, M. El-Batanouny, R. Sankar, and F. C. Chou. Anomalous behavior in the phonon dispersion of the (001) surface of Bi2Te3 determined from helium atom-surface scattering measurements. Phys. Rev. B. **88**, 035402 (2013).

Chapter 5
Pseudocharge Phonon Model

In order to identify the character and symmetry of the measured phonon events to be presented later, I employ empirical lattice dynamics calculations, based on the pseudocharge model (PCM) [1–3]. This model includes direct ion–ion interactions as well as indirect adiabatic coupling through the mediating electrons. Historically this model has been successful in reproducing surface phonon dispersion and HASS scattering amplitudes, derived from calculated surface charge deformations. Here I review some of the model's basic characteristics and tabulate the parameters used in our realization of the model.

5.1 Fundamentals of the Model

At the heart of the PCM is the expansion of the electron density n_l within each primitive cell l in terms of symmetry-adapted multipole components around selected Wyckoff symmetry points $\mathbf{R}_{lj}$.

$$n_l(\mathbf{r}) = \sum_{j\Gamma k} c_{\Gamma k}(lj) Y_{\Gamma k}(\mathbf{r} - \mathbf{R}_{lj}) \tag{5.1}$$

where Γ denotes an irreducible representation (irrep) of the Wyckoff symmetry point-group and k indexes its rows; $Y_{\Gamma k}$ is a symmetry-adapted harmonic basis function for the irrep Γ. The expansion coefficients $c_{\Gamma k}$ are separated into static and dynamic components, the latter being treated as bona-fide time-dependent dynamical variables,

$$c_{\Gamma k}(lj; t) = c_{\Gamma k}^{(0)}(lj) + \Delta c_{\Gamma k}(lj; t) \tag{5.2}$$

Note that the static components $c_{\Gamma k}^{(0)}(lj)$ at equivalent positions j will be identical, however their dynamical counterparts $\Delta c_{\Gamma k}(lj;t)$ will vary with both l and j.

Taylor expanding the potential energy to second order both in ionic displacements and pseudocharge (PC) deformations allows the Lagrangian of the combined pseudocharge-ion system to be written as

$$\mathcal{L} = \frac{1}{2}\left(\sum_{l\kappa\alpha} M_\kappa \dot{u}_\alpha^2(l\kappa) + \sum_{\substack{\Gamma k \\ lj}} m_\Gamma \Delta \dot{c}_{\Gamma k}^2(lj) - \left[\mathbf{u}\cdot\boldsymbol{\Phi}\cdot\mathbf{u} + (\mathbf{u}\cdot\mathbf{T}\cdot\Delta\mathbf{c} + \text{h.c.}) + \Delta\mathbf{c}\cdot\mathbf{H}\cdot\Delta\mathbf{c}\right]\right) \tag{5.3}$$

where $u_\alpha(l\kappa)$ is the displacement in the α direction of the ion at site κ in unit cell l; M is the ionic mass, and m_Γ is an effective PC mass that will be set to zero upon invoking the adiabatic approximation. $\boldsymbol{\Phi}$, $\mathbf{T}$, and $\mathbf{H}$ are empirical force-constant matrices representing ion–ion, ion–PC, and PC–PC interactions, respectively. The kinetic terms contain contributions from both the ions and pseudocharge (PC).

The Euler–Lagrange equations of motion for the ions and PCs can be obtained in the standard way, yielding

$$M_\kappa \ddot{u}_\alpha(l\kappa) = -\sum_{l'\kappa'\beta} \Phi_{\alpha\beta}\begin{pmatrix} l & l' \\ \kappa & \kappa' \end{pmatrix} u_\beta(l'\kappa') - \sum_{\substack{l'j \\ \Gamma k}} T_{\Gamma k}^\alpha \begin{pmatrix} l & l' \\ \kappa & j \end{pmatrix} \Delta c_{\Gamma k}(l'j) \tag{5.4}$$

$$m_\Gamma \Delta \ddot{c}_{\Gamma k}(lj) = -\sum_{l'\kappa'\alpha} T_{\Gamma k}^\alpha \begin{pmatrix} l & l' \\ j & \kappa' \end{pmatrix} u_\alpha(l'\kappa') - \sum_{l'j'} H_{\Gamma k}\begin{pmatrix} l & l' \\ j & j' \end{pmatrix} \Delta c_{\Gamma k}(l'j') \tag{5.5}$$

noting that only PCs belonging to the same irrep and same row can couple.

5.2 Adiatbatic Approximation, Ionic Self-Terms, and PC Self-Terms

First, I invoke the adiabatic approximation in which I set $m_\Gamma = 0$. This is equivalent to saying that the electronic response to lattice deformations is instantaneous, an approximation warranted by the factor of at least 10^3 in the ratio of the ionic and electron masses. Equation (5.5) then becomes

$$\Delta \mathbf{c} = -\mathbf{H}^{-1}\mathbf{T}^T \mathbf{u} \tag{5.6}$$

The entries of the matrices $\boldsymbol{\Phi}$ and $\mathbf{H}$ are expressed in terms of empirical parameters. However, the diagonal, self-term, elements that determine how the displacement of a particular ion affects its own motion must be treated carefully.

Next, the crystal must remain invariant under an arbitrary rigid displacement of all ions and PC by $\mathbf{u_0}$. In this case no ion experiences any acceleration and the left-hand side of (5.4) will be a null matrix of length $3N$ where N is the number of ions

5.2 Adiatbatic Approximation, Ionic Self-Terms, and PC Self-Terms

in the crystal. Combining this observation with Eq. (5.6) yields

$$0 = -\mathbf{\Phi}\mathbf{u}_0 + \mathbf{T}\mathbf{H}^{-1}\mathbf{T}^T\mathbf{u}_0 \tag{5.7}$$

Rearranging and separating the self-term from the rest of the sum gives

$$\Phi\begin{pmatrix} l\, l \\ \kappa\kappa \end{pmatrix} = -\sum_{l'\kappa'}' \Phi\begin{pmatrix} l\, l' \\ \kappa\kappa' \end{pmatrix} + \sum_{\substack{l'jj' \\ \Gamma k}} T_{\Gamma k}\begin{pmatrix} l\, l' \\ \kappa j \end{pmatrix} H_{\Gamma k}^{-1}\begin{pmatrix} l\, l' \\ jj' \end{pmatrix} T_{\Gamma k}^T\begin{pmatrix} l\, l' \\ \kappa j \end{pmatrix} \tag{5.8}$$

where the prime on the first sum indicates that the self-term is excluded. Equation (5.8) uniquely defines the ionic self terms to be consistent with translational invariance.

I proceed in a similar manner to calculate the diagonal elements of the matrix **H**. Again, I exploit translational invariance but this time explicitly choose the rigid displacement to be in the z-direction for clarity. Equation (5.5) can be written as

$$0 = -\sum_{l'\kappa'} T_{\Gamma k}\begin{pmatrix} l\, l' \\ j\, \kappa' \end{pmatrix}_z u_z(l'\kappa') - \sum_{l'j'} H_{\Gamma k}\begin{pmatrix} l\, l' \\ jj' \end{pmatrix}\Delta c_{\Gamma k}(l'j') \tag{5.9}$$

Separating the PC self-term from the rest of the sum and rearranging yields

$$H_{\Gamma k}\begin{pmatrix} l\, l \\ j\, j \end{pmatrix} = \frac{-1}{\Delta c_{\Gamma k}(l,j)} \left(\sum_{l'\kappa'} T_z\begin{pmatrix} l\, l' \\ j\, \kappa' \end{pmatrix} u_z(l',\kappa') + \sum_{l'j'}' H\begin{pmatrix} l\, l' \\ jj' \end{pmatrix}\Delta c_{\Gamma k}(l',j') \right) \tag{5.10}$$

For a rigid displacement of the entire crystal in the z-direction we have $\Delta c_{\Gamma k} = u_z = u_0$ for all l, κ, j. The PC self-term then takes the form

$$H_{\Gamma k}\begin{pmatrix} l\, l \\ j\, j \end{pmatrix} = -\sum_{l'\kappa'} T_z\begin{pmatrix} l\, l' \\ j\, \kappa' \end{pmatrix} - \sum_{l'j'}' H_{\Gamma k}\begin{pmatrix} l\, l' \\ j\, j' \end{pmatrix} \tag{5.11}$$

With the ionic and PC self-terms fixed I am now in a position to calculate the phonon frequencies. First, I make the simplifying substitutions

$$\tilde{\mathbf{u}}(l\kappa) = \sqrt{M_\kappa}\mathbf{u}(l\kappa)$$

$$\tilde{\Phi}\begin{pmatrix} l\, l' \\ \kappa\kappa' \end{pmatrix} = \frac{1}{\sqrt{M_\kappa M_\kappa'}} \Phi\begin{pmatrix} l\, l' \\ \kappa\kappa' \end{pmatrix}$$

$$\tilde{\mathbf{T}}\begin{pmatrix} l\, l' \\ \kappa j \end{pmatrix} = \frac{1}{\sqrt{M_\kappa}} \mathbf{T}\begin{pmatrix} l\, l' \\ \kappa j \end{pmatrix}$$

Then, Fourier transforming Eq. (5.4) and employing Eq. (5.6) yields

$$\omega^2(\mathbf{q})\tilde{\mathbf{u}}(\mathbf{q}) = \mathcal{D}(\mathbf{q})\tilde{\mathbf{u}}(\mathbf{q})$$
$$\mathcal{D}(\mathbf{q}) = \tilde{\boldsymbol{\Phi}}(\mathbf{q}) - \tilde{\mathbf{T}}(\mathbf{q})\mathbf{H}^{-1}(\mathbf{q})\tilde{\mathbf{T}}^\dagger(\mathbf{q}) \qquad (5.12)$$

Thus, to determine the phonon frequencies at a particular wave-vector one needs only construct the dynamical matrix $\mathcal{D}(\mathbf{q})$ and find its eigenvalues.

5.3 Bulk Parameters

The first step in any realization of the PCM is to populate the ionic positions. As for the PC, one is free to choose a high-symmetry point about which to expand the electron density. For this particular calculation, the "c" Wyckoff positions of the $R\bar{3}m$ space group, with C_{3v} point-group symmetry, were the most appropriate to use as centers of PC symmetry-adapted multipole expansion. They are identified as having coordinates $(0, 0, \pm z)$ that define the vertical axes of the tetrahedral pyramids shown in Fig. 5.2. The pyramid centers were chosen as PC expansion points. A comprehensive table of the Wyckoff positions for the $R\bar{3}m$ space group can be found in Fig. 5.1. C_{3v} has irreps A_1 (with dipolar symmetry-adapted harmonic z) and E (with dipolar symmetry-adapted harmonics x, y). In order to minimize the number of empirical constants employed in the bulk calculations, I opted to include only the A_1 symmetry-adapted fluctuations as depicted in Fig. 5.2, which shows the ion and PC locations in the PCM.

In the insulating bulk we do not include interactions between PCs, rendering **H** diagonal and constrained by Eq. (5.11). As for **T**, we introduce two force-constant parameters T_z^1 and T_z^2 to account for the ion-PC coupling in pyramids involving X1-Bi and X2-Bi, respectively. Finally, we use central ion–ion interaction potentials $v(r)$ with force-constant matrix elements of the form

$$\Phi_{\alpha\beta} = A\frac{x_\alpha x_\beta}{r_0^2} - B\left(\frac{x_\alpha x_\beta}{r_0^3} - \frac{1}{r_0}\delta_{\alpha\beta}\right) \qquad (5.13)$$

The parameters A and B are related to the ion potential via

$$A = \left.\frac{\partial^2 v}{\partial r^2}\right|_{r=r_0} \qquad B = \left.\frac{\partial v}{\partial r}\right|_{r=r_0} \qquad (5.14)$$

where r_0 is the equilibrium bond length.

As was mentioned earlier, the lack of available neutron scattering data for Bi_2Se_3 means the degree of agreement between experimental data and our calculation is determined only by the frequencies at the Γ point of the bulk BZ. I found it sufficient to include only nearest neighbor ion–ion interactions, except when it came

5.3 Bulk Parameters

Wyckoff Positions of Group 166 (R-3m) [h axes]

Multiplicity	Wyckoff letter	Site symmetry	Coordinates (0,0,0) + (2/3,1/3,1/3) + (1/3,2/3,2/3) +
36	i	1	(x,y,z) (-y,x-y,z) (-x+y,-x,z) (y,x,-z) (x-y,-y,-z) (-x,-x+y,-z) (-x,-y,-z) (y,-x+y,-z) (x-y,x,-z) (-y,-x,z) (-x+y,y,z) (x,x-y,z)
18	h	.m	(x,-x,z) (x,2x,z) (-2x,-x,z) (-x,x,-z) (2x,x,-z) (-x,-2x,-z)
18	g	.2	(x,0,1/2) (0,x,1/2) (-x,-x,1/2) (-x,0,1/2) (0,-x,1/2) (x,x,1/2)
18	f	.2	(x,0,0) (0,x,0) (-x,-x,0) (-x,0,0) (0,-x,0) (x,x,0)
9	e	.2/m	(1/2,0,0) (0,1/2,0) (1/2,1/2,0)
9	d	.2/m	(1/2,0,1/2) (0,1/2,1/2) (1/2,1/2,1/2)
6	c	3m	(0,0,z) (0,0,-z)
3	b	-3m	(0,0,1/2)
3	a	-3m	(0,0,0)

Fig. 5.1 Wyckoff positions of the $R\bar{3}m$ space group

Table 5.1 Bulk parameters for the Bi_2Se_3 lattice dynamical calculations based on the PCM

Ion–ion interaction			Ion-PC interaction	
Bond	A (N/m)	B (N)	Position	Value (N/m)
Se1–Se1	0.1	0.01	T_z^1 (Se1–Bi)	0.807
Se1–Bi	1.35	0.135	T_z^2 (Se2–Bi)	0.746
Se2–Bi	0.3	0.03		
Bi–Bi	0.2	0.02		

to the larger Bi atoms, where I admitted coupling to the other Bi atoms within the same layer, which are actually second nearest neighbors. The best fit to the experimental data is shown in Fig. 5.3. Dispersions are presented along three high-symmetry directions Λ (Γ-Z), Δ (Γ-X), and Σ (Γ-Y). Raman and IR data from [6] are presented as red and blue circles, respectively. The experimental parameters used in the calculation for Bi_2Se_3 are presented in Table 5.1.

In the case of Bi_2Te_3 the existence of available neutron scattering data allows one to perform a more rigorous fit, especially off the Γ point. In addition to the nominal intralayer Bi–Bi coupling employed in the case of Bi_2Se_3, I found it necessary to introduce an extra parameter coupling any given Bi atom to the other Bi layer within a single QL (interlayer). The force-constant and PC parameters were determined by

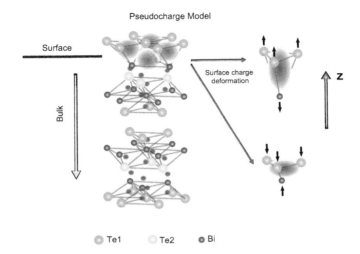

Fig. 5.2 Diagram of the ion and PC locations in the PCM. Although shown here for Bi_2Te_3, the structure is unchanged for the case of Bi_2Se_3. The figures to the *right* show the dipolar PC deformation associated with lattice distortions along the z direction. The *larger purple lobes* at the surface indicate the increased malleability of the surface PC, reflecting the presence of the metallic surface states. Figure from [8]

Table 5.2 Bulk parameters for the Bi_2Te_3 lattice dynamical calculations based on the PCM

Ion–ion interaction			Ion–PC interaction	
Bond	A (N/m)	B (N)	Position	Value (N/m)
Te1–Te1	0.187	0.0187	T_z^1 (Te1–Bi)	0.35
Te1–Bi	0.99	0.099	T_z^2 (Te2–Bi)	0.4
Te2–Bi	0.2	0.02		
Bi–Bi (intra)	0.2	0.02		
Bi–Bi (inter)	0.2	0.02		

fitting the bulk phonon calculation to available Raman, IR, and inelastic neutron spectroscopy data [4–7]. A summary of the parameters used and their values are given in Table 5.2.

The best fit to the available bulk data is shown in Fig. 5.4a–d. Dispersions are presented along three high-symmetry directions Λ (Γ-Z), Δ (Γ-X), and Σ (Γ-Y). Raman, IR, and neutron data are depicted as red triangles, green squares, and blue circles, respectively. The calculations agree well with all data types at the Γ point. However, there are some discrepancies between the calculations and the neutron data for some optical branches. First, there is disagreement in the dispersion of the low-energy E_u and E_g optical modes along the Λ-direction. Introduction of E PC deformations would not remedy this, since they involve interactions in the

5.3 Bulk Parameters

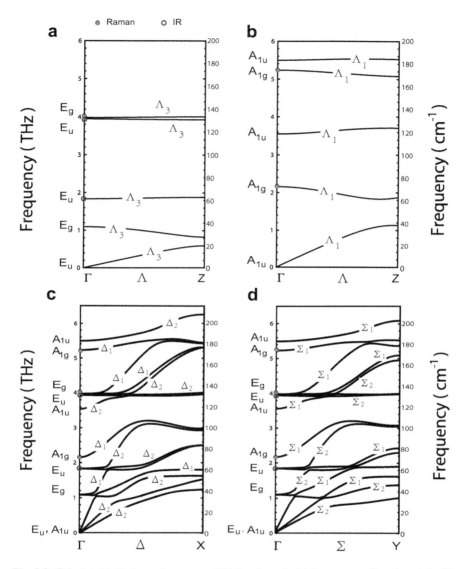

Fig. 5.3 Calculated bulk dispersion curves of Bi_2Se_3 along the high-symmetry directions Λ (**a–b**), Δ (**c**), and Σ (**d**). The C_{3v} symmetry of the Λ-direction allows one to project out purely longitudinal A modes and doubly degenerate, transverse E modes. Modes along the Δ and Σ directions have mixed polarization. The calculated dispersions fit the available Raman and IR data quite well. Figure from [9]

x/y plane and will not introduce the necessary phases along the Λ (z)-direction. I believe that the remedy for this discrepancy would be the introduction of long-range Coulomb interactions, as suggested by a previous study [5] of bulk Bi_2Te_3. However,

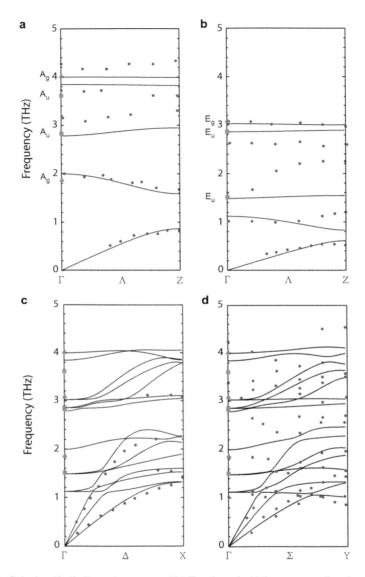

Fig. 5.4 Calculated bulk dispersion curves of Bi_2Te_3 along the high-symmetry directions Λ (**a–b**), Δ (**c**), and Σ (**d**). The calculated dispersions were fit to available Raman (*red triangles*), IR (*green squares*), and inelastic neutron scattering (*blue circles*) data. Figure from [8]

since I am primarily concerned with the surface, where Coulomb interactions are effectively screened by DFQs, I opted not to include such interactions. Second, there are some disagreements and ambiguities between the calculations and neutron data for the dispersions of high-energy optical modes along the Σ-direction. In this case the inclusion of E PC deformations might improve the fitting. Yet, I opted to omit these deformations since, as the reader will find in Chap. 6, I am primarily interested in the low-energy sector where a Kohn anomaly appears and there is already reasonable agreement.

5.4 Surface Parameters

Calculation of the surface phonon dispersions requires an adjustment to the nominal bulk geometry employed in our computational model. Instead of performing calculations for an infinitely periodic crystal in three dimensions I employ what is known as the *slab geometry*. The unit cell in this geometry becomes a finite alternating diatomic chain in the z direction created by stacking 30 QLs. Retention of periodicity in the x/y plane allows one to Fourier transform the equations of motion much like the case of Eq. (5.12). However, in this case calculations are performed along high-symmetry directions of the SBZ (Fig. 2.2) where the wave vectors are constrained to the x/y surface plane.

The surface of the material presents a unique environment for the constituent atoms and thus warrants an adjustment to the nominal bulk parameters described in Tables 5.1 and 5.2. This is especially true in the case of TIs where the surfaces also host metallic DFQ states. I again treat the surface parameters as empirical and adjust them to attain the best fit to the experimental HASS data presented in Chap. 6. Beyond the data, the following adjustments are also substantiated by physical reasoning:

1. The surface X1-Bi force-constant parameter was reduced to roughly 42 % of its bulk value to account for the reduced bonding and the emergence of metallic electrons.
2. A new planar (next-nearest neighbor) force-constant parameter involving intralayer surface X1–X1 bonds was introduced because of the reduced number of nearest neighbors when compared to their bulk counterparts. Furthermore, the metallic bonding occurring in the surface layer has a longer range than the insulating bulk.
3. Symmetry-adapted x/y deformations of the PC in the surface and subsurface pyramids, which form a basis of the doubly degenerate irrep E, were introduced to account for the delocalized nature of the DFQs. These are effected via new parameters T^S_{xy} and $\tilde{T}^S_{xy}$, respectively. In addition, T^S_z was reduced from its bulk value to account for the extra screening provided by the DFQ surface states.
4. A momentum dependent coupling H_q between dipolar z deformations of neighboring surface PC was introduced to account for interactions among the DFQs.

The surface parameters for Bi_2Te_3 are summarized in Table 5.3, with similar adjustments being made for Bi_2Se_3. The calculated dispersions for the slab geometry, along with accompanying HASS data, are presented in Chap. 6.

Table 5.3 Modified surface parameters for Bi_2Te_3

Surface ion–ion interaction			Surface ion–PC interaction	
Bond	A (N/m)	B (N)	Position	Value (N/m)
Te1–Bi	0.42	0.042	T_z^S (Te1–Bi)	0.24
Te1–Te1 (intra)	0.25	0.025	T_{xy}^S (Te1–Bi)	0.15
			$\tilde{T}_{xy}^S$ (Te2–Bi)	0.4

Surface PC–PC interaction

$H_q = H_0(1 + \frac{q^2}{a}e^{-q^2/b})$, $H_0 = -0.0782$, $a = 0.0034$, $b = 0.0075$

References

1. G. Benedek, M. Bernasconi, V. Chis, E. Chulkov, P.M. Echenique, B. Hellsing, J.P. Toennies, Theory of surface phonons at metal surfaces: recent advances. J. Phys. Condens. Matter **22**(8), 084020 (2010)
2. C.S. Jayanthi, H. Bilz, W. Kress, G. Benedek, Nature of surface-phonon anomalies in noble metals. Phys. Rev. Lett. **59**(7), 795–798 (1987)
3. C. Kaden, P. Ruggerone, J.P. Toennies, G. Zhang, G. Benedek, Electronic pseudocharge model for the Cu(111) longitudinal-surface-phonon anomaly observed by helium-atom scattering. Phys. Rev. B **46**(20), 13509–13525 (1992)
4. W. Kullmann, J. Geurts, W. Richter, N. Lehner, H. Rauh, U. Steigenberger, G. Eichhorn, R. Geick, Effect of hydrostatic and uniaxial pressure on structural properties and Raman active lattice vibrations in Bi2Te3. Phys. Status Solidi (B) **125**(1), 131–138 (1984)
5. W. Kullmann, G. Eichhorn, H. Rauh, R. Geick, G. Eckold, U. Steigenberger, Lattice dynamics and phonon dispersion in the narrow gap semiconductor Bi2Te3 with sandwich structure. Phys. Status Solidi (B) **162**(1), 125–140 (1990)
6. W. Richter, C.R. Becker, A raman and far-infrared investigation of phonons in the rhombohedral $V_2 - VI_3$ compounds Bi_2Te_3, Bi_2Se_3, Sb_2Te_3 and $Bi_2(Te_{1-x}Se_x)_3$ ($0 < x < 1$), $(Bi_{1-y}Sb_y)_2Te_3$ ($0 < y < 1$). Phys. Status Solidi (B) **84**(2), 619–628 (1977)
7. V. Wagner, G. Dolling, B.M. Powell, G. Landweher, Lattice vibrations of Bi2Te3. Phys. Status Solidi (B) **85**(1), 311–317 (1978)
8. C. Howard, M. El-Batanouny, R. Sankar, F.C. Chou, Anomalous behavior in the phonon dispersion of the (001) surface of Bi2Te3 determined from helium atom-surface scattering measurements. Phys. Rev. B **88**, 035402 (2013)
9. X. Zhu, L. Santos, R. Sankar, S. Chikara, C. Howard, F.C. Chou, C. Chamon, M. El-Batanouny, Interaction of phonons and Dirac fermions on the surface of Bi2Se3: a strong Kohn anomaly. Phys. Rev. Lett. **107**, 186102 (2011)

Chapter 6
HASS Results from the Surface of Bi_2Se_3 and Bi_2Te_3

This chapter contains the main experimental data presented in this thesis. I begin by presenting the elastic diffraction results that I used to determine the crystallographic orientation of the surface. From there, I move on to the measured inelastic data along with calculated surface phonon dispersion curves. Finally, I end this chapter with a calculation of the mode-specific electron–phonon coupling parameter $\lambda_\nu(\mathbf{q})$ based upon the experimental results.

6.1 Elastic and Inelastic Scattering Results

Typical diffraction patterns for both the $\bar{\Gamma}\bar{M}$ and $\bar{\Gamma}\bar{K}\bar{M}$ directions of Bi_2Te_3 are shown in Fig. 6.1. The raw measurements are collected as number of detector counts at different values of θ_f. In these plots θ_f has first been converted to ΔK using Eq. (3.8), which is then normalized by the magnitudes of the primitive reciprocal lattice vectors along the $\bar{\Gamma}\bar{M}$ and $\bar{\Gamma}\bar{K}\bar{M}$ directions, respectively. The results show well-defined elastic scattering peaks at integer values of the abscissa, indicating a clean, well-ordered surface in good agreement with the nominal lattice vectors. Similar results were obtained for Bi_2Se_3.

After each of these scans was taken, the azimuthal and polar controls of the manipulator were fixed and inelastic data was acquired by moving the detector around the sample and employing the TOF techniques described earlier. Typical TOF scans collected along different high-symmetry directions, after converting the abscissa from time to energy, are shown in Fig. 6.2. These scans constitute the raw data from which I calculated the surface phonon dispersion. Each scan was fit with the minimum number of Gaussian peaks necessary to fit the data. In practice the signal consisted of both elastic and inelastic peaks. The centers of the inelastic peaks define the energy transfer of the beam, and therefore the phonon energy. Employing

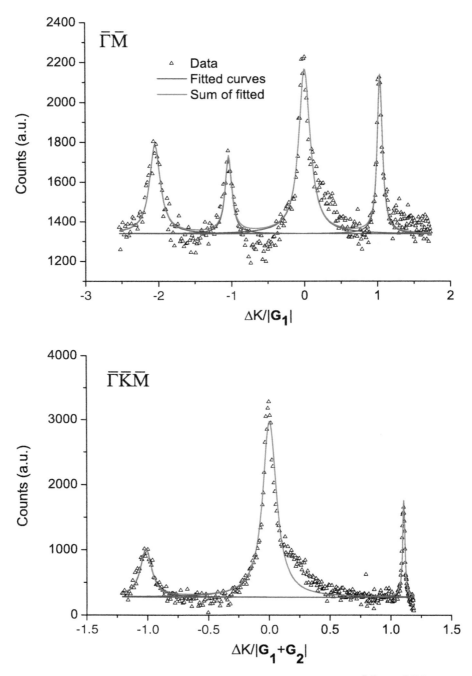

Fig. 6.1 Diffraction patterns indicating the two high-symmetry directions $\bar{\Gamma}\bar{M}$ and $\bar{\Gamma}\bar{K}\bar{M}$ on the surface of Bi_2Te_3. The *horizontal axis* depicts momentum transfer normalized to the pertinent primitive lattice vector. Figure from [13]

6.1 Elastic and Inelastic Scattering Results

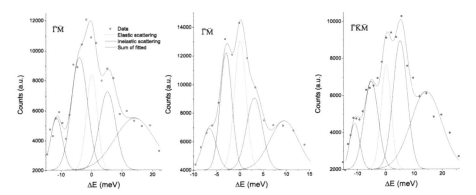

Fig. 6.2 Time of flights scans for the inelastic HASS measurements. The abscissa has been converted to energy difference for clarity. Phonon creation and annihilation events are manifest as *blue* inelastic peaks. Figure from [13]

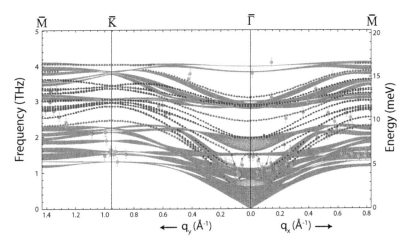

Fig. 6.3 Phonon dispersion along high-symmetry directions $\bar{\Gamma}\bar{K}\bar{M}$ and $\bar{\Gamma}\bar{M}$ for Bi_2Te_3. *Green and blue dots* indicate surface modes polarized perpendicular and parallel to the surface plane, respectively. The TOF measurements are depicted as *orange dots* with *error bars*. Figure from [13]

Eq. (3.10) allows one to determine the beam momentum transfer, and hence the phonon momentum q. Thus, each inelastic peak represents a single data point in the dispersion curves that follow.

The measured and computed dispersion curves for Bi_2Te_3 and Bi_2Se_3 are presented in Figs. 6.3 and 6.4, respectively. The gray areas are sets of projections of many distinct bulk phonon modes that form energy bands in the SBZ. The surface phonon modes were defined as those with at least 30 % of the oscillator strength (determined by the square of the mode eigenvector) concentrated within the first

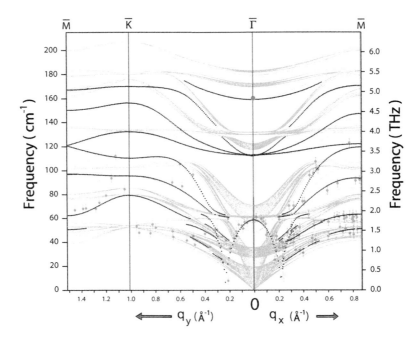

Fig. 6.4 Phonon dispersion for Bi_2Se_3 along the same high-symmetry directions as in Fig. 6.3. The *dotted lines* indicate surface phonon modes. The TOF measurements are depicted as *orange dots* with *error bars*. Figure from [12]

three atomic layers of the slab surface regions. There are two key features visible in both dispersions worth noting from the onset. First, there exists an optical surface phonon branch originating at approximately 1.4 THz for Bi_2Te_3 and 1.8 THz in the case of Bi_2Se_3 at the $\bar{\Gamma}$ point that disperses to lower energy with increasing wavevector in both the $\bar{\Gamma}\bar{M}$ and $\bar{\Gamma}\bar{K}$ directions. Calculations within the PCM confirm that these branches have vertical shear polarization with atomic displacements perpendicular to the surface layer. This trend terminates in V-shaped minima at $q \approx 0.08\,\text{Å}^{-1}$ in Bi_2Te_3 and $q \approx 0.2\,\text{Å}^{-1}$ in Bi_2Se_3, values that correspond to $2k_F$ for each crystal and thus signify the presence of Kohn anomalies [9].

In both cases, the isotropy of the optical phonon branch and its apparent termination at $2k_F$ can be explained by a scenario involving the DFQ surface states, in particular, their isotropic Fermi surface. In this scenario, the V-shaped feature marks the boundary between an operative DFQ screening for $q < 2k_F$ and its suppression above this value, which is a typical signature of a Kohn anomaly. Lattice dynamics calculations reveal some bulk penetration of vertical shear modes for $q > 2k_F$ reflecting a diminished role of DFQ screening and more compatibility with the insulating bulk. Indeed, scattering events with a momentum transfer greater than the diameter of the Fermi surface require energy, and are therefore suppressed. This is manifest as the recovery of the optical phonon branch dispersion after $q = 2k_F$.

6.1 Elastic and Inelastic Scattering Results

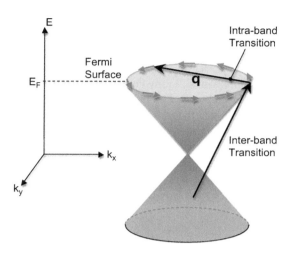

Fig. 6.5 Diagram showing the dispersion of the surface DFQs. The spin chirality of the Fermi surface is depicted by the *red arrows*. The screening effects of the DFQs can be understood in terms of the scattering of electrons in the Dirac cone with the momentum transfer supplied by a phonon. There are two distinct types of transitions: intra-band and inter-band, although the latter are suppressed by energetic considerations. Note that low-energy intra-band transitions are confined to a circle of diameter $2k_F$. Figure from [13]

The effective screening provided by scattering of the DFQs at the Fermi surface is depicted schematically in Fig. 6.5.

One should also note the absence of surface acoustic phonon modes in both the measured and computed dispersions of each material. Acoustic phonon modes with $q < 2k_F$ are absent from the dispersion whereas those with $q > 2k_F$ emerge in the form of a z-polarized Rayleigh mode beyond $2k_F$ for both materials. Three theoretical studies [6, 10, 11] considered the interaction of DFQs with long-wavelength surface acoustic modes. In all three the strength of the EPC was found to be quite small, which is actually consistent with my results. As a matter of fact, they justify the absence of acoustic Rayleigh phonons in HASS data: As was mentioned in Chap. 3, it is well established [2, 5] that the thermal energy helium atoms employed in HASS are scattered by the surface electron density about 2–3 Å above the terminal layer of atomic nuclei. Thus, detection of surface phonons by HASS involves scattering from the phonon-induced surface electron density oscillations. The results of the three theoretical studies confirm that surface acoustic phonons are weakly coupled to the surface metallic charge-density (DFQs) so that the induced density oscillations are effectively suppressed. I also point out that a recent study using density functional theory [8] has also demonstrated the absence of long wavelength Rayleigh modes in the phonon dispersion. However, one should note that the authors' results are for Bi_2Te_3 thin films (2–3 QLs) so the connection to the current slab geometry may be tenuous. Indeed, the study also yields a value for the average EPC constant that is significantly smaller than the result to be presented later.

One may question why there are several measured low-energy events that do not overlap with any computed surface phonon modes, especially in the case of Bi_2Te_3. Upon closer inspection, these events occur mainly near the SBZ boundaries where there is a large phonon density of states associated with a high concentration of flat, narrow projected bulk bands. It may be possible that these data points originate from inelastic events caused by exciting bulk phonon modes via surface resonances. Alternatively, it is possible that there exist surface phonon modes in those locations of the SBZ with an oscillator strength below the threshold used in the calculations.

6.2 Calculation of EPC Parameter in the Random Phase Approximation

In this section I establish the intimate link between the dispersive character of optical phonon branches exhibiting Kohn anomalies and the surface DFQ state response to ionic displacements. I then describe the phenomenological model-fitting approach applied to the experimentally measured dispersion of these optical phonon branches, and the procedure followed to extract the corresponding EPC function $\lambda_n(\mathbf{q})$. The construction of the model is carried out with the aid of the Random Phase Approximation (RPA).[1]

I start by defining the noninteracting, or free, surface phonons Hamiltonian in second-quantized form

$$\mathcal{H}_{ph} = \sum_{\mathbf{q},\nu} \hbar\omega_{\mathbf{q},\nu}^{(0)} \left(b_{\mathbf{q},\nu}^\dagger b_{\mathbf{q},\nu} + \frac{1}{2} \right) \quad (6.1)$$

where $b_{\mathbf{q},\nu}^\dagger$ is the creation operator of a phonon of bare frequency $\omega_{\mathbf{q},\nu}^{(0)}$ and branch index ν. The free phonon Matsubara Green's function of the $(\mathbf{q},\nu)$ mode is defined as

$$\mathcal{D}_\nu^{(0)}(\mathbf{q},i\omega_n) = \frac{2\left(\hbar\omega_{\mathbf{q},\nu}^{(0)}\right)}{(i\omega_n)^2 - \left(\hbar\omega_{\mathbf{q},\nu}^{(0)}\right)^2} \quad (6.2)$$

where $i\omega_n$ is the Matsubara "frequency," which actually has dimensions of energy.

The electronic surface states of Bi_2Te_3 form a two-dimensional Dirac metal, whose low-energy physics is well described by the Hamiltonian

$$\mathcal{H}_{el} = \sum_{\mathbf{k}} \psi_{\mathbf{k}}^\dagger \left[\hbar v_F \hat{\mathbf{z}} \cdot (\mathbf{k} \times \boldsymbol{\sigma}) - \mu\right] \psi_{\mathbf{k}} \quad (6.3)$$

---

[1]The RPA theoretical analysis, including derivation of the Dyson equation and Matsubara phonon Green function, was developed by Luiz Santos as part of a collaboration with the Chamon group at Boston University.

6.2 Calculation of EPC Parameter in the Random Phase Approximation

where $\psi_{\mathbf{k}} \equiv \begin{pmatrix} c_{\mathbf{k}\uparrow} \\ c_{\mathbf{k}\downarrow} \end{pmatrix}$ is the two-component electron spinor operator at wave-vector $\mathbf{k}$, v_F is the Fermi velocity, μ is the Fermi energy (which lies above the Dirac point), and $\boldsymbol{\sigma} = (\sigma_1, \sigma_2)$ is the vector containing the first two Pauli matrices. The Dirac Hamiltonian (6.3) is diagonal in the helicity basis $\Psi_{\mathbf{k}} = \begin{pmatrix} \gamma_{\mathbf{k}}^+ \\ \gamma_{\mathbf{k}}^- \end{pmatrix}$:

$$\Psi_{\mathbf{k}} = U_{\mathbf{k}} \psi_{\mathbf{k}}, \quad U_{\mathbf{k}} = \frac{1}{\sqrt{2}} \begin{pmatrix} i\, e^{i\varphi_{\mathbf{k}}} & 1 \\ -i\, e^{i\varphi_{\mathbf{k}}} & 1 \end{pmatrix}, \quad \varphi_{\mathbf{k}} \equiv \arctan\left(\frac{k_y}{k_x}\right) \quad (6.4)$$

yielding

$$\mathcal{H}_{\mathrm{el}} = \sum_{\mathbf{k}} \sum_{\alpha=\pm} \xi_{\mathbf{k}}^{\alpha} (\gamma_{\mathbf{k}}^{\alpha})^{\dagger} \gamma_{\mathbf{k}}^{\alpha}, \quad \xi_{\mathbf{k}}^{\alpha} = \alpha\, \hbar v_F |\mathbf{k}| - \mu \quad (6.5)$$

I consider an interaction between the electron density and the displacement $\mathbf{u}_j$ of the jth ion about its in-plane equilibrium position $\mathbf{R}_j^{(0)}$. The displacement $\mathbf{u}_j$ has both in-plane and out-of-plane components. The e–p interaction can be generically written as

$$\mathcal{H}_{\mathrm{el\text{-}ph}} = \int d^2\mathbf{r}\, \rho_{\mathrm{el}}(\mathbf{r}) \sum_j \boldsymbol{\eta}\left(\mathbf{r} - \mathbf{R}_j^{(0)}\right) \cdot \mathbf{u}_j \quad (6.6)$$

where $\rho_{\mathrm{el}}(\mathbf{r}) = \sum_{\sigma=\uparrow,\downarrow} c_{\sigma}^{\dagger}(\mathbf{r}) c_{\sigma}(\mathbf{r})$ is the electron surface density operator and $\boldsymbol{\eta}(\mathbf{r} - \mathbf{R}_j^{(0)})$ is a position dependent vector function (with units of energy per length) characterizing the EPC. The quantities ρ_{el}, $\boldsymbol{\eta}$, and $\mathbf{u}_j$ are then expanded in reciprocal-space as

$$\boldsymbol{\eta}\left(\mathbf{r} - \mathbf{R}_j^{(0)}\right) = \frac{1}{\mathcal{A}} \sum_{\mathbf{q}} \boldsymbol{\eta}_{\mathbf{q}}\, e^{i\mathbf{q}\cdot(\mathbf{r} - \mathbf{R}_j^{(0)})}$$

$$\rho_{\mathrm{el}}(\mathbf{r}) = \sum_{\sigma=\uparrow,\downarrow} c_{\sigma}^{\dagger}(\mathbf{r}) c_{\sigma}(\mathbf{r}) = \frac{1}{\mathcal{A}} \sum_{\sigma=\uparrow,\downarrow} \sum_{\mathbf{q}} e^{-i\mathbf{q}\cdot\mathbf{r}} \sum_{\mathbf{k}} c_{\mathbf{k}+\mathbf{q},\sigma}^{\dagger} c_{\mathbf{k},\sigma}$$

$$\mathbf{u}_j = \frac{1}{\sqrt{N}} \sum_{\mathbf{q},\nu} \sqrt{\frac{\hbar}{2M\omega_{\mathbf{q},\nu}^{(0)}}}\, e^{i\mathbf{q}\cdot\mathbf{R}_j^{(0)}} (b_{\mathbf{q},\nu} + b_{-\mathbf{q},\nu}^{\dagger})\, \hat{\mathbf{e}}_{\nu}(\mathbf{q})$$

where N is the number of surface primitive cells, $\mathcal{A}$ is the surface area, and $\hat{\mathbf{e}}_{\nu}(\mathbf{q})$ is the polarization vector. Substitution in (6.6) leads to the e–p interaction Hamiltonian

$$\mathcal{H}_{\mathrm{el\text{-}ph}} = \frac{1}{\sqrt{\mathcal{A}}} \sum_{\sigma=\uparrow,\downarrow} \sum_{\mathbf{k}} \sum_{\mathbf{q},\nu} g_{\mathbf{q},\nu}\, c_{\mathbf{k}+\mathbf{q},\sigma}^{\dagger} c_{\mathbf{k},\sigma} (b_{\mathbf{q},\nu} + b_{-\mathbf{q},\nu}^{\dagger}) \quad (6.7)$$

with the e–p coupling

$$g_{q,\nu} = \sqrt{\frac{N\hbar}{2MA\,\omega_{q,\nu}^{(0)}}}\,\eta_q \cdot \hat{e}_\nu(q) \equiv \sqrt{\frac{N\hbar}{2MA\,\omega_{q,\nu}^{(0)}}}\,\eta_{q,\nu} \quad (6.8)$$

The dressed Matsubara phonon Green function $\mathcal{D}_\nu$ is obtained from the following diagrammatic equation

$$\mathcal{D}_\nu = \mathcal{D}_\nu^{(0)} + \mathcal{D}_\nu^{(0)} \frac{|g_\nu|^2}{\varepsilon} \Pi \mathcal{D}_\nu \quad (6.9)$$

where $|g_\nu|^2/\varepsilon$ represents the vertex interactions with ε the dielectric function, and the RPA bubble is the electron polarization function defined as

$$\Pi(\mathbf{q}, i\omega_n) = \frac{1}{\mathcal{A}} \frac{1}{\beta} \sum_{i\Omega_m} \sum_{\mathbf{p}} \mathrm{Tr}\left[G^{(0)}_{\sigma,\sigma'}(\mathbf{p}+\mathbf{q}, i\Omega_m + i\omega_n) G^{(0)}_{\sigma,\sigma'}(\mathbf{p}, i\Omega_m) \right]$$
(6.10)

where $\mathcal{A}$ is the surface area, and Tr acts on the spin degrees of freedom $\sigma, \sigma' = \uparrow, \downarrow$. $G^{(0)}$ is the noninteracting electronic Matsubara Green's function with fermionic Matsubara frequencies Ω_m

$$G^{(0)}_{\sigma,\sigma'}(\mathbf{p}, i\Omega_m) = -\int_0^\beta d\tau\, e^{i\Omega_m \tau} \left\langle T_\tau\, c_{\mathbf{p},\sigma}(\tau)\, c^\dagger_{\mathbf{p},\sigma'}(0) \right\rangle_0 \quad (6.11)$$

where $\beta \equiv 1/k_B T$ and T_τ is the imaginary time-ordering operator. Performing the Matsubara sums in Eq. (6.10) yields

$$\Pi(\mathbf{q}, i\omega_n) = \int \frac{d^2\mathbf{k}}{(2\pi)^2} \frac{1+\cos\theta}{2} \left[\frac{n_F(\xi^+_{\mathbf{k}+\mathbf{q}}) - n_F(\xi^+_\mathbf{k})}{\xi^+_{\mathbf{k}+\mathbf{q}} - \xi^+_\mathbf{k} - i\omega_n} + \frac{n_F(\xi^-_{\mathbf{k}+\mathbf{q}}) - n_F(\xi^-_\mathbf{k})}{\xi^-_{\mathbf{k}+\mathbf{q}} - \xi^-_\mathbf{k} - i\omega_n} \right]$$
(6.12)

where ξ^+ and ξ^- were defined in Eq. (6.5), n_F is the Fermi function, and θ is the angle between wave vectors $\mathbf{k}$ and $\mathbf{k} + \mathbf{q}$. The factor $(1 + \cos\theta)$ accounts for the effect of DFQ chirality introduced by strong spin–orbit coupling. As such, I note here that the contribution of the spin-orbit interaction to EPC is manifest explicitly in the polarization, or response, function as well as implicitly in the strong vertex

6.2 Calculation of EPC Parameter in the Random Phase Approximation

interactions. In contrast, a recent study [4] of the EPC in ultrathin Bi films using density functional perturbation theory reports that the spin-orbit coupling mainly operates through the vertex terms rather than through the response function. The difference perhaps arises from the fact that, in TIs like Bi_2Te_3, the very existence of DFQs and their chirality is a direct manifestation of spin-orbit coupling, which is clearly reflected in the response function.

The RPA dielectric function is given by

$$\varepsilon(\mathbf{q}, i\omega_n) = 1 - v_c(\mathbf{q}) \Pi(\mathbf{q}, i\omega_n) \tag{6.13}$$

where $v_c(\mathbf{q}) = \frac{e^2}{2\varepsilon_0 |\mathbf{q}|}$ is the two-dimensional Fourier transform of the electron–electron Coulomb interaction potential. Solving the diagrammatic equation (6.9) yields

$$\mathcal{D}_\nu(\mathbf{q}, i\omega_n) = \frac{\mathcal{D}_\nu^{(0)}(\mathbf{q}, i\omega_n)}{1 - \mathcal{D}_\nu^{(0)}(\mathbf{q}, i\omega_n) |g_{\mathbf{q},\nu}|^2 \frac{\Pi(\mathbf{q}, i\omega_n)}{\varepsilon(\mathbf{q}, i\omega_n)}} = \frac{2(\hbar \omega_{\mathbf{q},\nu}^{(0)})}{(i\omega_n)^2 - (\hbar \omega_{\mathbf{q},\nu}^{(0)})^2 - 2(\hbar \omega_{\mathbf{q},\nu}^{(0)})\tilde{\Pi}} \tag{6.14}$$

with $\tilde{\Pi} = |g_{\mathbf{q},\nu}|^2 \frac{\Pi(\mathbf{q}, i\omega_n)}{\varepsilon(\mathbf{q}, i\omega_n)}$ being the phonon self-energy.

After performing the analytic continuation $i\omega_n \to \omega + i0^+$, I obtain the renormalized frequencies as the real part of the poles of $\mathcal{D}_\nu(\mathbf{q}, \omega)$

$$(\hbar \omega_{\mathbf{q},\nu})^2 = (\hbar \omega_{\mathbf{q},\nu}^{(0)})^2 + 2(\hbar \omega_{\mathbf{q},\nu}^{(0)}) \operatorname{Re}\left[\tilde{\Pi}(\mathbf{q}, \omega_{\mathbf{q},\nu})\right] \tag{6.15}$$

$\operatorname{Re}\left[\tilde{\Pi}(\mathbf{q}, \omega_{\mathbf{q},\nu})\right]$ is then adjusted to reproduce the measured phonon dispersion. It depends on the two parameters $\eta_\parallel$ and $\eta_\perp$ that appear in the coupling function $\eta_{\mathbf{q},\nu}$, which lie in the sagittal-plane with directions parallel and normal to the wave-vector $\mathbf{q}$, respectively. In view of the near constancy of the ionic screened potential $V(\mathbf{q})$ for $q \leq 2k_F$, and the fact that the electron–phonon coupling involves the gradient of a screened potential, I write

$$\eta_{\mathbf{q},\nu} = \eta_\perp + \frac{q}{2k_F}\eta_\parallel \tag{6.16}$$

A more detailed definition of these couplings can be found elsewhere [12]. The values of the bare phonon frequency $\omega^{(0)}$ and k_F are extracted from experimental results. The former is identified as the experimental value of $\omega(\mathbf{q}=0)$, where the DFQ response vanishes, while $2k_F$ was set as the wave-vector where the V-shaped Kohn anomaly occurs. A summary of these parameters for both Bi_2Te_3 and Bi_2Se_3 can be found in Table 6.1.

Table 6.1 Parameters used in the RPA calculation

	Bi_2Te_3	Bi_2Se_3
$\eta_\parallel$ (eV Å)	108	65
$\eta_\perp$ (eV Å)	71	50
$\omega^{(0)}$ (THz)	1.4	1.8
k_F (Å^{-1})	0.04	0.1

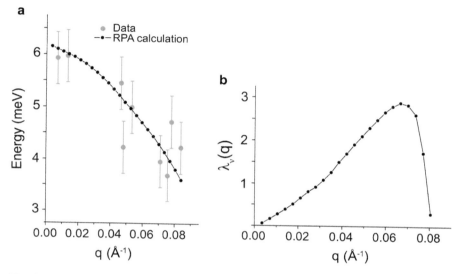

Fig. 6.6 Panel (**a**) depicts the experimental data along both high-symmetry directions in the SBZ collapsed to a single axis along with the RPA calculation of the dressed phonon frequencies. The EPC function is depicted in panel (**b**). Note the increase in coupling with increasing wave vector and sudden termination at $2k_F$ where the DFQ response vanishes. Figure from [13]

After fitting $\text{Re}\left[\tilde{\Pi}(\mathbf{q}, \omega_{\mathbf{q},\nu})\right]$ to the experimental dispersion curve, the corresponding $\text{Im}\left[\tilde{\Pi}(\mathbf{q}, \omega_{\mathbf{q},\nu})\right]$ is obtained by a Kramers–Kronig transformation

$$\text{Im}\left[\tilde{\Pi}(\mathbf{q}, \omega_{\mathbf{q}})\right] = \frac{2}{\pi} \int_0^\infty \frac{\omega_{\mathbf{q}}}{\omega_{\mathbf{q}}^2 - \omega_{\mathbf{q}}'^2} \text{Re}\left[\tilde{\Pi}(\mathbf{q}, \omega_{\mathbf{q}}')\right] d\omega_{\mathbf{q}}' \quad (6.17)$$

Finally, the EPC function is obtained from the relation [1, 3, 7]

$$\lambda_\nu(\mathbf{q}) = -\frac{\text{Im}[\tilde{\Pi}(\mathbf{q}, \omega_{\mathbf{q},\nu})]}{\pi \mathcal{N}(E_F)(\hbar\omega_{\mathbf{q},\nu})^2} \quad (6.18)$$

where $\mathcal{N}(E_F)$ is the density of electronic states at the Fermi surface. The mode-specific EPC function for Bi_2Te_3 is shown in Fig. 6.6, with similar results holding for Bi_2Se_3. One can average over the wave vector and branch index dependence of the EPC function to obtain

$$\bar{\lambda} = \sum_\nu \frac{1}{\pi(2k_F)^2} \int_0^{2k_F} \int_0^{2\pi} \lambda_\nu(\mathbf{q}) \, q dq \, d\phi = \sum_\nu \frac{1}{2k_F^2} \int_0^{2k_F} \lambda_\nu(q) \, q dq \qquad (6.19)$$

which is useful when comparing to values in the literature determined from electron spectroscopy measurements, where the information regarding ν and $\mathbf{q}$ is lost. In the present case the data depicts only one phonon branch experiencing renormalization, so the sum contains a single term. Performing this integral for each material I find $\bar{\lambda}^{Te} \approx 2$ and $\bar{\lambda}^{Se} \approx 0.7$.

References

1. P.B. Allen, Neutron spectroscopy of superconductors. Phys. Rev. B **6**, 2577–2579 (1972)
2. G. Benedek, M. Bernasconi, K.P. Bohnen, D. Campi, E. Chulkov, P.M. Echenique, R. Heid, I.Y. Sklyadneva, J.P. Toennies, Unveiling mode-selected electron–phonon interactions in metal films by helium atom scattering. Phys. Chem. Chem. Phys. **16**, 7159–7172 (2014)
3. W.H. Butler, F.J. Pinski, P.B. Allen, Phonon linewidths and electron–phonon interaction in Nb. Phys. Rev. B **19**, 3708–3721 (1979)
4. V. Chis, G. Benedek, P.M. Echenique, E.V. Chulkov, Phonons in ultrathin Bi(1 1 1) films: role of spin-orbit coupling in electron–phonon interaction. Phys. Rev. B **87**, 075412 (2013)
5. D. Farias, K.-H. Rieder, Atomic beam diffraction from solid surfaces. Rep. Prog. Phys. **61**(12), 1575 (1998)
6. S. Giraud, R. Egger, Electron–phonon scattering in topological insulators. Phys. Rev. B **83**, 245322 (2011)
7. G. Grimvall, *The Electron–Phonon Interaction in Metals* (North-Holland, Amsterdam, 1981)
8. G.Q. Huang, Surface lattice vibration and electron–phonon interaction in topological insulator Bi2Te3 (111) films from first principles. Europhys. Lett. **100**(1), 17001 (2012)
9. W. Kohn, Image of the fermi surface in the vibration spectrum of a metal. Phys. Rev. Lett. **2**, 393–394 (1959)
10. S.D. Sarma, Q. Li, Many-body effects and possible superconductivity in the two-dimensional metallic surface states of three-dimensional topological insulators. Phys. Rev. B **88**, 081404 (2013)
11. P. Thalmeier, Surface phonon propagation in topological insulators. Phys. Rev. B **83**(12), 125314 (2011)
12. X. Zhu, L. Santos, R. Sankar, S. Chikara, C. Howard, F.C. Chou, C. Chamon, M. El-Batanouny, Interaction of phonons and dirac fermions on the surface of Bi_2Se_3: a strong Kohn anomaly. Phys. Rev. Lett. **107**, 186102 (2011)
13. C. Howard, M. El-Batanouny, R. Sankar, F.C. Chou, Anomalous behavior in the phonon dispersion of the (001) surface of Bi2Te3 determined from helium atom-surface scattering measurements. Phys. Rev. B **88**, 035402 (2013)

Chapter 7
Translating Between Electron and Phonon Perspectives

Having analyzed and quantified the DFQ–phonon interaction from the phonon perspective, I will now turn to the electron perspective. In this chapter I will show that the interaction Hamiltonian in Eq. (6.7) modifies the DFQ energies and lifetimes much like we have already seen in the case of the surface phonons. The matrix elements $g_{q,\nu}$ of the interaction Hamiltonian along with the dressed phonon propagator, both of which have already been determined from the phonon data, can be used as input to a Matsubara Green function formalism to calculate the modifications to the electron dispersion. I will present calculations of the DFQ spectral function for Bi_2Te_3, confirm the self-consistency of the approach, and discuss implications of the translation process.

7.1 Motivation

In studying DFQ-phonon coupling on the surfaces of topological insulators one can choose to probe either the electronic or vibrational (phonon) states and look for signatures of said coupling in the obtained spectra. Whereas the phonon spectra are usually probed using HASS or EELS, electronic eigenstates are traditionally probed via ARPES. Recent ARPES experiments have yielded widely varying estimates of the electron–phonon coupling parameter $\bar{\lambda}$ on the surfaces of topological insulators [1–4]. Without a way to connect the results originating from different spectroscopies, it is difficult to come to a consensus on the correct value. In the following sections I show that it is possible to connect the results of these experiments by determining the DFQ spectral function using information from measured phonon spectra. Signatures of the coupling present in the surface phonon dispersion curves can then be directly traced to modifications to the DFQ spectral function close to the Fermi energy. Determining $\bar{\lambda}$ from the calculated electron spectral function yields values consistent with the phonon spectroscopy results presented in Chap. 6 as well as recent high resolution ARPES data [4]. This novel methodology allows for effective translation between existing experimental methods and should usher in a consensus about the magnitude of the EPC on surfaces of topological insulators. Moreover, the formalism is not exclusive to phonons and can be used to examine other bosonic couplings in a variety of condensed matter systems.

7.2 DFQ Self-Energy Formalism

The DFQ–phonon interaction described in Eq. (6.7) modifies the DFQ propagator. It can be determined by evaluating the following expression

$$G(\mathbf{k}, \tau, T) = -\frac{1}{\mathcal{A}} \sum_{\substack{\mathbf{k}_1 \mathbf{k}_2 \\ \mathbf{q}}} |g_\mathbf{q}|^2 \iint_0^\beta d\tau_1 \, d\tau_2 \, \mathscr{D}(\mathbf{q}, \tau_1 - \tau_2)$$

$$\times \left\langle T_\tau \left[c^\dagger_{\mathbf{k}_1+\mathbf{q}}(\tau_1) c^\dagger_{\mathbf{k}_2-\mathbf{q}}(\tau_2) c_{\mathbf{k}_2}(\tau_2) c_{\mathbf{k}_1}(\tau_1) c_\mathbf{k}(\tau) c^\dagger_\mathbf{k}(0) \right] \right\rangle \quad (7.1)$$

where $\mathcal{A}$ is the surface area, $g_\mathbf{q}$ is the electron–phonon matrix element of the optical phonon branch identified in Chap. 6, $\beta \equiv 1/k_B T$, $\mathscr{D}(\mathbf{q}, \tau_1 - \tau_2)$ is the phonon propagator, $c^\dagger_\mathbf{k}$, $c_\mathbf{k}$ are the electron creation and annihilation operators, respectively, and T_τ is the imaginary time-ordering operator. In the following analysis I neglect the weak direct Coulomb interactions in the DFQ system on the TI surface. This is warranted by the fact that TIs possess large dielectric constants ($\kappa > 50$) and Fermi velocities $\sim 10^5$ m/s that yield a small effective fine-structure constant $\alpha = e^2/(\kappa \hbar v_F) \approx 0.05$ [5, 6]. Moreover, from a Fermi liquid perspective, the quasiparticle nature of the DFQs close to the Fermi energy (E_F) is well defined because of their substantially long lifetimes. Since the analysis will focus on a region ± 7 meV about E_F, the direct electron–electron interactions need not be considered.

Fourier transforming the Matsubara function gives the DFQ self-energy

$$\Sigma(\mathbf{k}, i\omega_n, T) = \frac{1}{\mathcal{A}} \sum_\mathbf{q} |g_\mathbf{q}|^2 \left(1 + \hat{\mathbf{k}} \cdot \widehat{\mathbf{k}+\mathbf{q}} \right)$$

$$\times \left[\frac{n_B(\omega_\mathbf{q}) + n_F(\varepsilon_{\mathbf{k}+\mathbf{q}})}{i\omega_n - \varepsilon_{\mathbf{k}+\mathbf{q}} + \omega_\mathbf{q}} + \frac{n_B(\omega_\mathbf{q}) + 1 - n_F(\varepsilon_{\mathbf{k}+\mathbf{q}})}{i\omega_n - \varepsilon_{\mathbf{k}+\mathbf{q}} - \omega_\mathbf{q}} \right] \quad (7.2)$$

where ω_n is the Matsubara "frequency," $\omega_\mathbf{q}$ is the optical phonon energy at wavevector $\mathbf{q}$, $\varepsilon_\mathbf{k}$ is the bare DFQ energy at wave vector $\mathbf{k}$, n_B is the Bose–Einstein distribution, and n_F is the Fermi–Dirac distribution. Full details of the derivation of Eq. (7.2) can be found in Appendix A. To get a qualitative sense of what this function embodies it is worth briefly examining each of the terms, which can be interpreted in terms of virtual and real phonon emission processes. Examining the first term, then

7.2 DFQ Self-Energy Formalism

at zero temperature the numerator is only finite if $|\mathbf{k}+\mathbf{q}| < k_F$, so the intermediate state is a hole. The pole in the first term occurs at $\varepsilon_{\mathbf{k}+\mathbf{q}} - \omega_\mathbf{q}$, corresponding to a state of one hole less one phonon. Thus, one may interpret this term as the energy shift that results from the adsorption of virtual phonons by holes. Moving to the second term, one sees that the numerator is only finite if the intermediate electron state is empty, i.e. $|\mathbf{k}+\mathbf{q}| > k_F$. Furthermore, the poles of the second expression are located at energies $\varepsilon_{\mathbf{k}+\mathbf{q}} + \omega_\mathbf{q}$, which is the energy of an electron of momentum $\mathbf{k}+\mathbf{q}$ and an emitted phonon of momentum $\mathbf{q}$, so the second process corresponds to phonon emission by an electron.

Moving forward, I shall use the dispersion curve and phonon matrix elements

$$\omega_\mathbf{q} = B - C\left(\frac{q}{k_F}\right)^2$$

$$|g_\mathbf{q}|^2 = D\left(1 + F\frac{q}{k_F}\right) \tag{7.3}$$

where $k_F = 0.04\,\text{Å}^{-1}$ is the Fermi wavevector, $B = 6.01\,\text{meV}$, $C = 0.55\,\text{meV}$, $D = 8.08 \times 10^5\,\text{meV}^2\text{Å}^2$, and $F = 1.52$. These values were obtained by fitting the HASS data of the renormalized optical phonon branch in Bi_2Te_3. Analytically continuing and replacing the sum by an integral, one obtains

$$\Sigma(\mathbf{k}, \omega, T) = \frac{1}{(2\pi)^2} \int d\mathbf{q}\, |g_\mathbf{q}|^2 \left(1 + \hat{\mathbf{k}} \cdot \widehat{\mathbf{k}+\mathbf{q}}\right)$$
$$\times \left[\frac{n_B(\omega_\mathbf{q}) + n_F(\varepsilon_{\mathbf{k}+\mathbf{q}})}{\omega - \varepsilon_{\mathbf{k}+\mathbf{q}} + \omega_\mathbf{q} + i\eta} + \frac{n_B(\omega_\mathbf{q}) + 1 - n_F(\varepsilon_{\mathbf{k}+\mathbf{q}})}{\omega - \varepsilon_{\mathbf{k}+\mathbf{q}} - \omega_\mathbf{q} + i\eta}\right] \tag{7.4}$$

with the imaginary part given by

$$\text{Im}[\Sigma(\mathbf{k}, \omega, T)] = -\frac{1}{4\pi} \int dq\, q \int d\varphi\, |g_\mathbf{q}|^2 \left(1 + \hat{\mathbf{k}} \cdot \widehat{\mathbf{k}+\mathbf{q}}\right)$$
$$\times \Big[\big(n_B(\omega_\mathbf{q}) + n_F(\varepsilon_{\mathbf{k}+\mathbf{q}})\big)\delta(\omega - \varepsilon_{\mathbf{k}+\mathbf{q}} + \omega_\mathbf{q})$$
$$+ \big(n_B(\omega_\mathbf{q}) + 1 - n_F(\varepsilon_{\mathbf{k}+\mathbf{q}})\big)\delta(\omega - \varepsilon_{\mathbf{k}+\mathbf{q}} - \omega_\mathbf{q})\Big] \tag{7.5}$$

I first determine $\text{Im}[\Sigma(\mathbf{k}, \omega, T)]$ numerically (see Appendix B) and then obtain $\text{Re}[\Sigma(\mathbf{k}, \omega, T)]$ via the Kramers–Kronig relations. Finally, I obtain the DFQ spectral function as

$$A(\mathbf{k}, \omega, T) = \frac{1}{\pi} \frac{|\text{Im}[\Sigma(\mathbf{k}, \omega, T)]|}{(\omega - \hbar v_0(|\mathbf{k}| - k_F) - \text{Re}[\Sigma(\mathbf{k}, \omega, T)])^2 + \text{Im}[\Sigma(\mathbf{k}, \omega, T)]^2} \tag{7.6}$$

where ω is measured from $E_F = \hbar v_0 k_F$ and I have used $\varepsilon_\mathbf{k} = \hbar v_0(|\mathbf{k}| - k_F)$ for the nominal dispersion of the DFQs above the Dirac point. Here, $v_0 = 4 \times 10^5$ m/s is the Fermi velocity in the absence of interactions. The determination of $A(\mathbf{k}, \omega, T)$ allows direct comparison of the quasiparticle energy dispersion and state broadening with experimental results obtained by ARPES measurements. The Lorentzian nature of the spectral function underscores the notion that $\text{Im}\left[\Sigma(\mathbf{k}, \omega, T)\right]$ modifies the lifetime of the quasiparticle states (as evidenced by the finite line width) while $\text{Re}\left[\Sigma(\mathbf{k}, \omega, T)\right]$ shifts the dispersion by modifying the poles of the expression. Moreover, the density of DFQ states per unit area

$$\mathcal{N}(\omega, T) = \frac{1}{(2\pi)^2} \int d\mathbf{k} A(\mathbf{k}, \omega, T) \quad (7.7)$$

allows for comparison with experimental results of ARPES energy distribution curves (EDCs) and scanning tunneling spectroscopy (STS).

7.3 Computational Results

For a fixed temperature, $\Sigma(\mathbf{k}, \omega, T)$ is a function of energy and wave vector. Thus, in order to evaluate this function I set up a fine computational grid in this 2D space and calculated $\text{Im}\left[\Sigma(\mathbf{k}, \omega, T)\right]$ at every point on the grid, resulting in a smooth surface. I then performed a Kramers Kronig transformation at every value of energy and wave vector to calculate the surface corresponding to $\text{Re}\left[\Sigma(\mathbf{k}, \omega, T)\right]$. The results for $T = 0.01 E_F \equiv T_1$ are shown in Fig. 7.1. This procedure was repeated for several different temperatures. Results at all temperatures indicate that both components of the DFQ self-energy depend only mildly on wave vector, with the most interesting texture occurring along the energy axis. For this reason, I have taken line cuts at a specific wave vector values to talk quantitatively about the function's structure. Those for T_1 and $T = 0.04 E_F \equiv T_2$ are shown in the left and right halves of Fig. 7.2, respectively. Noting that the interacting optical phonon band extends from

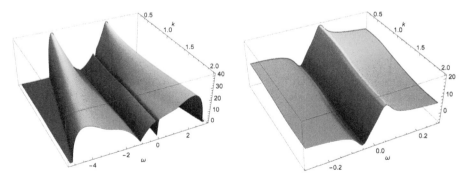

Fig. 7.1 Imaginary (*left*) and real (*right*) components of the DFQ self-energy for $T = 0.01 E_F$. The wave vector and energy axes have been scaled by k_F and E_F, respectively

7.3 Computational Results

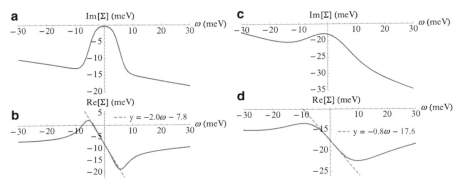

Fig. 7.2 Imaginary and real parts of the DFQ self-energy for Bi_2Te_3 at $T = 0.01E_F$ (**a**, **b**) and $T = 0.04E_F$ (**c**, **d**). The *dashed red lines* indicate a linear fit about $\omega = 0$ meV, whose slope is used to determine λ via Eq. (7.8). Figure from [11]

3–6.01 meV, the imaginary part experiences a dip in magnitude about E_F reflecting the enhanced lifetime and hence diminished line width of electronic states with $|\omega|$ less than the lower band edge. Note that increasing temperature softens this dip by increasing the population of electrons and holes above and below E_F, respectively. The real part displays the characteristic asymmetric variation across the Fermi level that effectively shifts the dispersion via Eq. (7.6). One will notice that the variation in the shift is attenuated at increased temperature as evidenced by the softer peaks in the real part.

Since reported high-resolution ARPES results were performed at temperatures in the range $T \sim 7\text{--}20$ K, I shall focus my analysis at the relatively low temperature T_1. With the real and imaginary parts of the self-energy at hand, I present the calculated spectral function for T_1 in Fig. 7.3a where the DFQ band dispersion appears as the bright yellow curve. The EPC footprint is readily apparent as deviation from the linear dispersive behavior within ± 7 meV of E_F, where two kinks appear, one slightly above and the other below E_F, pointing to large velocity renormalization. Note that the structural details of the kinks are discernible on an energy scale < 1 meV. This energy scale is an order of magnitude smaller than the ARPES resolutions used in two studies [1, 2], which could account for the fact that no such deviation in the dispersion was observed in those experiments. The application of higher resolution in more recent experiments [3, 4] brought some of these features to light.

Figure 7.3b shows $\Delta k(\omega)$, the full-width-half-maximum (FWHM) of the momentum distribution curves (MDCs), as the white region. Note that $\Delta k(\omega)$ increases from $0.2k_F$ at $\omega = -30$ meV to $0.28k_F$ at $\omega = -7$ meV, where the lower kink occurs, indicating a gradual increase in the strength of the EPC. The important observation is that $\Delta k(\omega)$ abruptly shrinks, reaching negligible values above $\omega = -2$ meV, and resumes a linear dispersion but with a slower velocity. In Fig. 7.3a/b this is manifest as dotted features that are the result of infinitesimal,

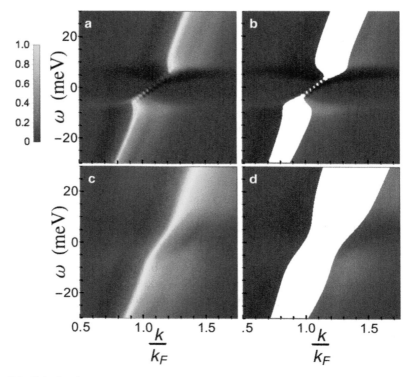

Fig. 7.3 Calculated spectral function and momentum distribution curve FWHM $\Delta k(\omega)$ for Bi_2Te_3 at T_1 (**a, b**) and T_2 (**c, d**). The *white space* in panels (**b, d**) is used to indicate $\Delta k(\omega)$. Figure from [11]

or delta-function-like, line widths that follow the discrete computational grid. The collapse in the peak width signals the absence of DFQ coupling to phonons. This termination of the coupling is consistent with the observation that the DFQs interact strongly with low-lying optical phonon modes, whose lower band-edge occurs at 3 meV, as reported in HASS data. Unlike acoustic phonons, the interaction does not extend to infinitesimal energies close to E_F, which is why $\Delta k(\omega)$ is suppressed in the region $\sim \pm 2$ meV.

I am now in position to determine $\bar{\lambda}$ from the electron perspective in two distinct, albeit equivalent ways. In the first method, I apply the definition [7, 8]

$$\bar{\lambda} = -\frac{\partial \text{Re}[\Sigma]}{\partial \omega}\bigg|_{\omega = E_F} \tag{7.8}$$

that has traditionally been used for metallic surfaces. The dashed red lines in Fig. 7.2 show a linear fit to $\text{Re}[\Sigma]$ at E_F, allowing us to determine $\bar{\lambda}$ from the slope. The alternative approach utilizes the relation

7.3 Computational Results

$$\bar{\lambda} = \frac{v_0}{v_F} - 1 \qquad (7.9)$$

where v_F is the EPC renormalized Fermi velocity obtained from the slope of our dispersion curves at E_F and v_0 is the un-renormalized value. Both definitions yield $\bar{\lambda} = 2$, which is consistent with results from phonon spectroscopy. As a matter of fact the two definitions can be connected as follows: The dispersion curve is defined as the maximum in $A(\mathbf{k}, \omega)$ at fixed ω. Equation (7.6) indicates that this maximum occurs when

$$\omega - \varepsilon(\mathbf{k}) - \mathrm{Re}\,[\Sigma] = 0 \qquad (7.10)$$

Here, we made use of the fact that $\mathrm{Im}\,[\Sigma]$ has a very weak dependence on $\mathbf{k}$, as it always does. Now, we write

$$\begin{aligned}
\left.\frac{\partial \omega}{\partial \mathbf{k}}\right|_{\omega=E_F} &= \mathbf{v}_F \\
&= \left.\left(\frac{\partial \varepsilon}{\partial \mathbf{k}} + \frac{\partial \mathrm{Re}\,[\Sigma]}{\partial \omega} \frac{\partial \omega}{\partial \mathbf{k}}\right)\right|_{\omega=E_F} \\
&= \mathbf{v}_0 - \bar{\lambda} \mathbf{v}_F
\end{aligned} \qquad (7.11)$$

which produces Eq. (7.9).

This expression evokes an analogy between the relativistic DFQs propagating on a TI surface and light traveling in a dielectric medium, which helps illuminate the physical meaning of $\bar{\lambda}$. First, we note that the nominal linear dispersion of the DFQs reflects their massless character, invalidating the application of the conventional idea of mass enhancement as defined by $m^*/m = 1 + \bar{\lambda}$ to TIs. Instead, it is appropriate to interpret $\bar{\lambda}$ as a velocity renormalization factor via $v_0/v_F = 1 + \bar{\lambda}$. We note that this is totally consistent with, and actually more fundamental than the previous relation, to which it simplifies when applied to parabolic energy bands. Thus, $\bar{\lambda}$ provides a measure of the renormalization of the group velocity of the DFQs near the Fermi energy, much like the index of refraction does for light in a dielectric medium. Just as light slows down when propagating through matter, the DFQs near the Fermi energy are slowed by their interactions with the phonon gas.

At this point it is appropriate to present the effects of increased temperature on the spectral function at T_2. The results are plotted in Fig. 7.3c/d. One can discern dramatic differences between the spectral functions at the two temperatures. First, the T_2 spectral function displays enhanced line broadening as compared with the T_1 spectral function. More importantly, it also exhibits a much smaller deviation from the nominal linear dispersion compared to its T_1 counterpart, emphasizing the fact that high resolution and cryogenic temperatures are required for adequate observation of the EPC manifest features in ARPES measurements. Indeed, I find a significantly higher v_F at T_2, yielding a markedly lower value of $\bar{\lambda}$ when applying Eq. (7.9). In Fig. 7.4 I present the calculated values of $\bar{\lambda}$ at several temperatures, including T_1 and T_2. Note that the strong variation of $\bar{\lambda}$ with temperature casts doubt on the applicability of a linear interpolation of the temperature dependence

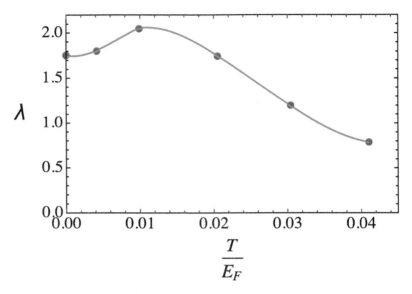

Fig. 7.4 The temperature dependence of the electron phonon parameter. After initially increasing in the region $0 \leq \frac{T}{E_F} \leq 0.01$, the value of $\bar{\lambda}$ begins to fall with temperature. Figure from [11]

of $\text{Im}[\Sigma(\omega = E_F, T)]$ to extract an estimate of $\bar{\lambda}$ via the relation $\text{Im}[\Sigma(\omega = E_F, T)] = \pi \bar{\lambda} k_B T$ [7]. This method assumes $\bar{\lambda}$ is constant over the temperature range of interest, which Fig. 7.4 shows is clearly not the case.

Another noteworthy feature in my results is the DFQ density of states and its derivative, which I present for T_1 in Fig. 7.5. Manifestations of the electron–phonon interactions are clearly seen when comparing the density of states in the absence of interactions, depicted by the dashed red line, with that in the presence of interactions, shown as a blue curve, in Fig. 7.5a. Moreover, when the density of states is multiplied by the Fermi function a peak-dip-hump structure appears below E_F, in agreement with recent EDC spectra [4]. Additionally, a dip-peak-dip-peak in $d\mathcal{N}/d\omega$ appear at -8 meV, -2 meV, 0 meV, and $+3$ meV, respectively, which are consistent with d^2I/dV^2 STS measurements by Madhavan's group (Private communication). The readily apparent texture in the density of states calculation suggests that STS is a valuable tool for observing signatures of EPC on the surfaces of TIs.

7.4 Additional Supporting Results

In the previous section I provided a detailed mathematical framework allowing me to translate between phonon and DFQ spectra for the particular case of Bi_2Te_3.

7.4 Additional Supporting Results

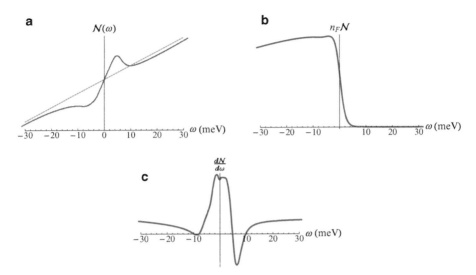

Fig. 7.5 (a) The density of states per unit area $\mathcal{N}(\omega)$ obtained from the calculated spectral function at T_1. The *dashed red line* depicts the non-interacting density of states. (b) $\mathcal{N}(\omega)$ multiplied by the Fermi function for comparison with EDCs measured by ARPES. (c) $d\mathcal{N}/d\omega$ for comparison with d^2I/dV^2 STS spectra. Figure from [11]

Calculating $\bar{\lambda}$ from the DFQ perspective via Eqs. (7.8) and (7.9) yielded results consistent with the calculation from the phonon perspective presented in Chap. 6, and hence demonstrated the self-consistency of the approach. Indeed, this is guaranteed by the fact that we used the same phonon dispersion $\omega_{\mathbf{q}}$ and matrix elements $g_{\mathbf{q},\nu}$ extracted from the surface phonon measurements. It is crucial to compare my results with those obtained *experimentally* to see if there is actually agreement between my calculations and experimental evidence from the DFQ perspective. I have already mentioned two experiments performed on Bi_2Te_3, one using ARPES [4] and another using STS (Private communication), that show agreement with our calculations of the spectral function, energy distribution curves, $\bar{\lambda}$, and density of states curves.

However it is worth noting that there are two additional experiments performed on Bi_2Se_3 that also agree with my experimental evidence from Chap. 6. The first came from an experiment [9] that showed a thermally activated coupling in transport measurements performed on the Bi_2Se_3 surface. The authors demonstrate that their results indicate a coupling to a single optical phonon of frequency ≈ 1.90 THz. Additionally, an even more recent experiment [10] employing time-resolved ARPES has provided evidence of coupling to a surface optical phonon of with frequency ≈ 2.05 THz. Both of these values agree with my measurement of 1.80 THz for the optical phonon frequency of Bi_2Se_3 at the $\bar{\Gamma}$ point in the SBZ when experimental uncertainty is taken into account. Thus, over the past 2 years there have emerged several high-resolution experiments probing the surfaces of these TIs from the electron perspective and yielding results consistent with my HASS measurements.

References

1. Z.-H. Pan, A.V. Fedorov, D. Gardner, Y.S. Lee, S. Chu, T. Valla, Measurement of an exceptionally weak electron–phonon coupling on the surface of the topological insulator Bi_2Se_3 using angle-resolved photoemission spectroscopy. Phys. Rev. Lett. **108**, 187001 (2012)
2. R.C. Hatch, M. Bianchi, D. Guan, S. Bao, J. Mi, B.B. Iversen, L. Nilsson, L. Hornekær, P. Hofmann, Stability of the **Bi_2Se_3**(111) topological state: electron–phonon and electron-defect scattering. Phys. Rev. B **83**, 241303 (2011)
3. C. Chen, Z. Xie, Y. Feng, H. Yi, A. Liang, S. He, D. Mou, J. He, Y. Peng, X. Liu, Y. Liu, L. Zhao, G. Liu, X. Dong, J. Zhang, L. Yu, X. Wang, Q. Peng, Z. Wang, S. Zhang, F. Yang, C. Chen, Z. Xu, X.J. Zhou, Tunable Dirac fermion dynamics in topological insulators. Sci. Rep. **3**, 08 (2013)
4. T. Kondo, Y. Nakashima, Y. Ota, Y. Ishida, W. Malaeb, K. Okazaki, S. Shin, M. Kriener, S. Sasaki, K. Segawa, Y. Ando, Anomalous dressing of dirac fermions in the topological surface state of Bi_2Se_3, Bi_2Te_3, and Cu-doped Bi_2Se_3. Phys. Rev. Lett. **110**, 217601 (2013)
5. S.D. Sarma, Q. Li, Many-body effects and possible superconductivity in the two-dimensional metallic surface states of three-dimensional topological insulators. Phys. Rev. B **88**, 081404 (2013)
6. S. Raghu, S.B. Chung, X.-L. Qi, S.-C. Zhang, Collective modes of a helical liquid. Phys. Rev. Lett. **104**, 116401 (2010)
7. G. Grimvall, *The Electron–Phonon Interaction in Metals* (North-Holland Publishing Company, Amsterdam, 1981)
8. P. Hofmann, I.Y. Sklyadneva, E.D.L. Rienks, E.V. Chulkov, Electron–phonon coupling at surfaces and interfaces. New J. Phys. **11**(12), 125005 (2009)
9. M.V. Costache, I. Neumann, J.F. Sierra, V. Marinova, M.M. Gospodinov, S. Roche, S.O. Valenzuela. Fingerprints of inelastic transport at the surface of the topological insulator Bi_2Se_3: role of electron–phonon coupling. Phys. Rev. Lett. **112**, 086601 (2014)
10. J.A. Sobota, S.-L. Yang, D. Leuenberger, A.F. Kemper, J.G. Analytis, I.R. Fisher, P.S. Kirchmann, T.P. Devereaux, Z.-X. Shen, Distinguishing bulk and surface electron–phonon coupling in the topological insulator Bi_2Se_3 using time-resolved photoemission spectroscopy. Phys. Rev. Lett. **113**, 157401 (2014)
11. C. Howard, M. El-Batanouny. Connecting electron and phonon spectroscopy data to consistently determine quasiparticle-phonon coupling on the surface of topological insulators. Phys. Rev. B **89**, 075425 (2014)

Chapter 8
Conclusion and Future Directions

Having presented the totality of my research and findings, I will use this chapter to conclude my dissertation. First, I will present a brief summary of my work and discuss its implications. Then I will move on to discuss the future research direction for the Boston University Surface Laboratory.

8.1 Summary

The main experimental achievement of my research was measuring the surface phonon dispersion of both Bi_2Se_3 and Bi_2Te_3 using HASS. In order to determine the symmetry and polarization of the measured data, I used the PCM to fit the experimental data. This procedure yielded a comprehensive exposition of the surface phonon dispersion along the high-symmetry directions in the SBZ. Two fascinating anomalies, common to both materials, were discovered in the surface phonon dispersion. First, the ubiquitous Rayleigh surface acoustic wave is absent in the experimental data and PCM calculations of the surface phonon dispersions for both Bi_2Se_3 and Bi_2Te_3. This experimental observation points to a potentially attractive device application for topological insulators. Specifically, this property in topological insulators appears to inhibit interfacial sound wave transmission, they could potentially be used in devices designed for soundproofing. Moreover, the surface phonon dispersions of both Bi_2Se_3 and Bi_2Te_3 exhibit a z-polarized optical mode centered at the $\bar{\Gamma}$ point that disperses to lower energy with increasing wave vector along both high-symmetry directions of the SBZ. This mode softening terminates in a V-shaped minimum at a wave vector corresponding to $2k_F$, constituting a Kohn anomaly, which I have attributed to the interaction between the surface phonons and DFQs. This evidence is crucial for understanding the coupling between the vibrational and electronic degrees of freedom at the surfaces of topological insulators and assessing their viability for applications in spintronics and quantum computing.

In addition to my experimental work I was able to quantify the aforementioned coupling by calculating the mode-specific DFQ-phonon coupling parameter $\lambda(\mathbf{q})$ using a phenomenological model based within the random phase approximation. This constitutes the first time that such detailed information has been extracted from phonon dispersion measurements. I also successfully translated the signatures of EPC present in my measured phonon spectra for Bi_2Te_3 (namely, the prominent low-lying optical modes and the associated Kohn anomaly) to the electron perspective using a Matsubara Green function framework. In doing so, I showed how anomalies in the surface phonon dispersion originating from EPC can be translated into modifications of the DFQ dispersion about the Fermi level. My results indicate that the signatures of the EPC occur on an energy scale of about 1 meV and are strongly temperature dependent. These findings set the necessary experimental resolutions and appropriate methods for quantifying the EPC from the electron perspective. Calculations of the averaged EPC parameter $\bar{\lambda}$ from both the phonon and electron perspectives consistently yield the same results, underscoring the reliability of my approach. Perhaps most importantly, the translation process demonstrated in this dissertation provides a method for connecting the results of phonon and photoemission spectroscopy experiments, which should help usher a consensus on the true value of $\bar{\lambda}$. Finally, the Matsubara formalism used in the translation process is quite general and be extended to other bosonic couplings besides phonons.

8.2 Future Work

The metallic surface states of TIs are guaranteed to exist provided TRS is unbroken and, as I have shown in this dissertation, are involved in a plethora of interesting physics. However, TIs are not the only materials that exhibit symmetry-protected Dirac-like states at their surfaces. As was mentioned in the introduction, initial theoretical predictions and subsequent experimental verification have shown that there exists a different class of topologically non-trivial insulators whose metallic surface states are protected by mirror symmetry rather than TRS. These materials have been coined *topological crystalline insulators* or TCIs. One set of potentially interesting candidates are the alloys $Pb_{1-x}Sn_xX$ (X=Se,Te), whose transition from the trivial to topological regime can be driven by both temperature and Sn concentration. A schematic of the band inversion is shown in Fig. 8.1, which depicts the evolution of the bulk bandgap with varying Sn concentration.

The ability to control the presence of the DFQs on the surface via a crystal growth technique is quite attractive. For one, it is far simpler than breaking TRS in the case of TIs and allows one to study the physics when DFQs are both present and absent

8.2 Future Work

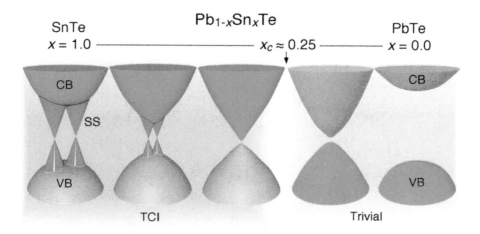

Fig. 8.1 Diagram showing that the appearance of Dirac surface states in $Pb_{1-x}Sn_xX$ can be controlled by varying the Sn concentration. Image from [1]

from the surface. Additionally, although the mechanism is still not understood, temperature appears to play an important role in driving the band inversion in $Pb_{1-x}Sn_xSe$ and therefore can also be used to control the appearance of the surface states. Figure 8.2 shows the interplay between temperature and Sn concentration in controlling the band inversion.

Crystalline $Pb_{1-x}Sn_xSe$ adopts the rocksalt structure and therefore possesses a square SBZ. Unlike the case of TIs, the TCI $Pb_{1-x}Sn_xSe$ possesses an even number (four) of Dirac cones in the SBZ. Each cone is offset slightly from the X point and therefore the surface electronic structure is comprised of many *double Dirac cones*. As in the case of TIs, these Dirac cones also possess spin texture.

Above and below the Dirac point the double Dirac cones hybridize to create a more complicated structure. This includes the presence of a saddle point, which leads to a dramatic change in the Fermi surface. To illustrate this, Fig. 8.3 shows the evolution of the Fermi surface as the chemical potential is tuned upward relative to the Dirac point.

One can see that for energies close to the Dirac point, the two cones are independent and the Fermi surface consists of two distinct circles. However, upon crossing the saddle point, the Fermi surface transforms into an ovular shape with a hole pocket inside. This abrupt change in the Fermi surface topology is known as a Lifshitz transition and could lead to interesting physics on these surfaces. Lastly, there is recent evidence [3] that a spontaneous lattice distortion on the surface breaks reduces the mirror symmetry and gaps out one of the Dirac cone pairs while preserving the other.

I have already shown that the Dirac surface states in the TIs Bi_2Se_3 and Bi_2Te_3 couple strongly to the surface phonons and lead to anomalies in their dispersion. In TCIs there is an even richer electronic structure with multiple double Dirac cones, a

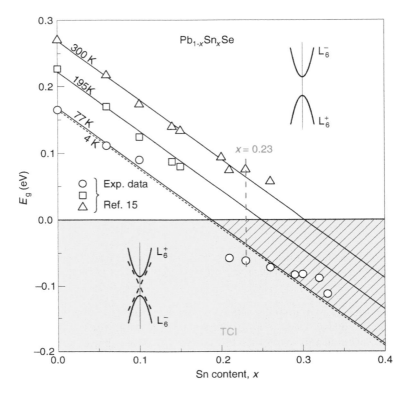

Fig. 8.2 Plot showing how the band inversion in $Pb_{1-x}Sn_xSe$ is affected by temperature and Sn concentration. Positive values of the gap energy E_g indicate the trivial phase while negative values indicate the topological phase. Image from [2]

Lifshitz transition, and a surface lattice distortion. An interesting and worthwhile endeavor for the Boston University Surface Laboratory would be to perform measurements of the surface phonon dispersion of $Pb_{1-x}Sn_xSe$ looking for unique signatures of EPC and how they change when modifying the surface electronic environment. In particular one could control the presence of the DFQs with temperature or Sn concentration and examine how the surface phonon dispersion changes as a result. Additionally the Lifshitz transition creates the possibility of changing scattering pathways by manipulating the required momentum transfer of the phonon as depicted in Fig. 8.4. Perhaps it is possible to map anomalies in the phonon dispersion at a particular wave vector **q** to particular electronic transitions and seeing how those anomalies change when varying the chemical potential.

At this point it should be obvious that both TIs and TCIs are host to fascinating metallic surface states whose presence is dictated by topology rather than order. This fact makes them particularly robust in the face of certain perturbations and therefore makes them attractive for device applications. However a thorough understanding of these Dirac states and their interaction with the ion dynamics is necessary

8.2 Future Work

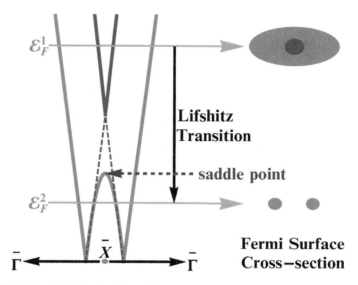

Fig. 8.3 Evolution of the Fermi surface with changing chemical potential. The pronounced change in the surface topology upon crossing the saddle point indicates a Lifshitz transition

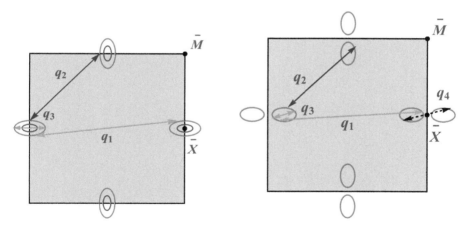

Fig. 8.4 Diagram depicting how the change in Fermi surface topology depicted in Fig. 8.3 can affect the phonon scattering processes

before they can be made technologically useful. This will require a combination of theoretical modeling, computation, and experimental evidence provided from both the phonon and electron perspectives. Only then can we hope to move topological materials from the laboratory to the marketplace.

References

1. Y. Tanaka, T. Sato, K. Nakayama, S. Souma, T. Takahashi, Z. Ren, M. Novak, K. Segawa, Y. Ando, Tunability of the k-space location of the dirac cones in the topological crystalline insulator $Pb_{1-x}Sn_xTe$. Phys. Rev. B **87**, 155105 (2013)
2. P. Dziawa, B.J. Kowalski, K. Dybko, R. Buczko, A. Szczerbakow, M. Szot, E. Łusakowska, T. Balasubramanian, B.M. Wojek, M.H. Berntsen, O. Tjernberg, T. Story, Topological crystalline insulator states in $Pb_{1-x}Sn_xSe$. Nat. Mat. **11**(12), 1023–1027, 12 (2012)
3. Y. Okada, M. Serbyn, H. Lin, D. Walkup, W. Zhou, C. Dhital, M. Neupane, S. Xu, Y.J. Wang, R. Sankar, F. Chou, A. Bansil, M.Z. Hasan, S.D. Wilson, L. Fu, V. Madhavan, Observation of Dirac node formation and mass acquisition in a topological crystalline insulator. Science **341**(6153), 1496–1499 (2013)

Appendix A
Supplemental Material for Electron Self-Energy Analysis

A.1 Electron Green's Function

The process of going from Eqs. (7.1)–(7.2) is not obvious and requires a thorough explanation which I present here. Beginning with the expression for the dressed electron propagator

$$G(\mathbf{k}, \tau, T) = -\frac{1}{\mathcal{A}} \sum_{\substack{\mathbf{k}_1 \mathbf{k}_2 \\ \mathbf{q}}} |g_\mathbf{q}|^2 \int_0^\beta d\tau_1 \int_0^\beta d\tau_2 \, \mathscr{D}(\mathbf{q}, \tau_1 - \tau_2)$$
$$\times \left\langle T_\tau \left[c^\dagger_{\mathbf{k}_1+\mathbf{q}}(\tau_1) \, c^\dagger_{\mathbf{k}_2-\mathbf{q}}(\tau_2) \, c_{\mathbf{k}_2}(\tau_2) \, c_{\mathbf{k}_1}(\tau_1) \, c_\mathbf{k}(\tau) \, c^\dagger_\mathbf{k}(0) \right] \right\rangle \quad \text{(A.1)}$$

one can apply Wick's theorem and the canonical commutation relations for the electron creation and annihilation operators to reorder them in the vacuum expectation value at the expense of additional terms. This yields

$$G(\mathbf{k}, \tau, T) = \frac{1}{\mathcal{A}} \sum_{\substack{\mathbf{k}_1 \mathbf{k}_2 \\ \mathbf{q}}} |g_\mathbf{q}|^2 \int_0^\beta d\tau_1 \int_0^\beta d\tau_2 \, \mathscr{D}(\mathbf{q}, \tau_1 - \tau_2)$$
$$\times \Big\langle T_\tau \Big[c_{\mathbf{k}_2}(\tau_2) \, c^\dagger_{\mathbf{k}_1+\mathbf{q}}(\tau_1) \, c_{\mathbf{k}_1}(\tau_1) \, c^\dagger_{\mathbf{k}_2-\mathbf{q}}(\tau_2) \, c_\mathbf{k}(\tau) \, c^\dagger_\mathbf{k}(0)$$
$$- c_{\mathbf{k}_1}(\tau_1) \, c^\dagger_{\mathbf{k}_1+\mathbf{q}}(\tau_1) \, c_{\mathbf{k}_2}(\tau_2) \, c^\dagger_{\mathbf{k}_2-\mathbf{q}}(\tau_2) \, c_\mathbf{k}(\tau) \, c^\dagger_\mathbf{k}(0)$$
$$+ c_{\mathbf{k}_1}(\tau_1) \, c^\dagger_{\mathbf{k}_1+\mathbf{q}}(\tau_1) \, c_\mathbf{k}(\tau) \, c^\dagger_{\mathbf{k}_2-\mathbf{q}}(\tau_2) \, c_{\mathbf{k}_2}(\tau_2) \, c^\dagger_\mathbf{k}(0)$$

$$
\begin{aligned}
&- c_{\mathbf{k}_2}(\tau_2) c^\dagger_{\mathbf{k}_1+\mathbf{q}}(\tau_1) c_{\mathbf{k}}(\tau) c^\dagger_{\mathbf{k}_2-\mathbf{q}}(\tau_2) c_{\mathbf{k}_1}(\tau_1) c^\dagger_{\mathbf{k}}(0) \\
&+ c_{\mathbf{k}}(\tau) c^\dagger_{\mathbf{k}_1+\mathbf{q}}(\tau_1) c_{\mathbf{k}_2}(\tau_2) c^\dagger_{\mathbf{k}_2-\mathbf{q}}(\tau_2) c_{\mathbf{k}_1}(\tau_1) c^\dagger_{\mathbf{k}}(0) \\
&- c_{\mathbf{k}}(\tau) c^\dagger_{\mathbf{k}_1+\mathbf{q}}(\tau_1) c_{\mathbf{k}_1}(\tau_1) c^\dagger_{\mathbf{k}_2-\mathbf{q}}(\tau_2) c_{\mathbf{k}_2}(\tau_2) c^\dagger_{\mathbf{k}}(0) \Big] \Big\rangle
\end{aligned} \quad \text{(A.2)}
$$

Now one can express the vacuum expectation value as a sum of products of non-interacting electron Greens functions.

$$
\begin{aligned}
G(\mathbf{k}, \tau, T) = &\frac{1}{\mathcal{A}} \sum_{\substack{\mathbf{k}_1 \mathbf{k}_2 \\ \mathbf{q}}} |g_\mathbf{q}|^2 \int_0^\beta d\tau_1 \int_0^\beta d\tau_2 \, \mathscr{D}(\mathbf{q}, \tau_1 - \tau_2) \\
&\times \Big[G^{(0)}_{\mathbf{k}_2, \mathbf{k}_1+\mathbf{q}}(\tau_2 - \tau_1) G^{(0)}_{\mathbf{k}_1, \mathbf{k}_2-\mathbf{q}}(\tau_1 - \tau_2) G^{(0)}_\mathbf{k}(\tau) \\
&- G^{(0)}_{\mathbf{k}_1, \mathbf{k}_1+\mathbf{q}}(0) G^{(0)}_{\mathbf{k}_2, \mathbf{k}_2-\mathbf{q}}(0) G^{(0)}_\mathbf{k}(\tau) \\
&+ G^{(0)}_{\mathbf{k}_1, \mathbf{k}_1+\mathbf{q}}(0) G^{(0)}_{\mathbf{k}, \mathbf{k}_2-\mathbf{q}}(\tau - \tau_2) G^{(0)}_{\mathbf{k}_2, \mathbf{k}}(\tau_2) \\
&- G^{(0)}_{\mathbf{k}_2, \mathbf{k}_1+\mathbf{q}}(\tau_2 - \tau_1) G^{(0)}_{\mathbf{k}, \mathbf{k}_2-\mathbf{q}}(\tau - \tau_2) G^{(0)}_{\mathbf{k}_1, \mathbf{k}}(\tau_1) \\
&+ G^{(0)}_{\mathbf{k}, \mathbf{k}_1+\mathbf{q}}(\tau - \tau_1) G^{(0)}_{\mathbf{k}_2, \mathbf{k}_2-\mathbf{q}}(0) G^{(0)}_{\mathbf{k}_1, \mathbf{k}}(\tau_1) \\
&- G^{(0)}_{\mathbf{k}, \mathbf{k}_1+\mathbf{q}}(\tau - \tau_1) G^{(0)}_{\mathbf{k}_1, \mathbf{k}_2-\mathbf{q}}(\tau_1 - \tau_2) G^{(0)}_{\mathbf{k}_2, \mathbf{k}}(\tau_2) \Big]
\end{aligned} \quad \text{(A.3)}
$$

At this point one can perform a Fourier transform in imaginary time to yield

$$
\begin{aligned}
G(\mathbf{k}, i\omega_n, T) = &\frac{1}{\mathcal{A}} \sum_{\substack{\mathbf{k}_1 \mathbf{k}_2 \\ \mathbf{q}}} |g_\mathbf{q}|^2 \int_0^\beta d\tau \, e^{i\omega_n \tau} \int_0^\beta d\tau_1 \int_0^\beta d\tau_2 \, \mathscr{D}(\mathbf{q}, \tau_1 - \tau_2) \\
&\times \Big[G^{(0)}_{\mathbf{k}_2, \mathbf{k}_1+\mathbf{q}}(\tau_2 - \tau_1) G^{(0)}_{\mathbf{k}_1, \mathbf{k}_2-\mathbf{q}}(\tau_1 - \tau_2) G^{(0)}_\mathbf{k}(\tau) \\
&- G^{(0)}_{\mathbf{k}_1, \mathbf{k}_1+\mathbf{q}}(0) G^{(0)}_{\mathbf{k}_2, \mathbf{k}_2-\mathbf{q}}(0) G^{(0)}_\mathbf{k}(\tau) \\
&+ G^{(0)}_{\mathbf{k}_1, \mathbf{k}_1+\mathbf{q}}(0) G^{(0)}_{\mathbf{k}, \mathbf{k}_2-\mathbf{q}}(\tau - \tau_2) G^{(0)}_{\mathbf{k}_2, \mathbf{k}}(\tau_2) \\
&- G^{(0)}_{\mathbf{k}_2, \mathbf{k}_1+\mathbf{q}}(\tau_2 - \tau_1) G^{(0)}_{\mathbf{k}, \mathbf{k}_2-\mathbf{q}}(\tau - \tau_2) G^{(0)}_{\mathbf{k}_1, \mathbf{k}}(\tau_1) \\
&+ G^{(0)}_{\mathbf{k}, \mathbf{k}_1+\mathbf{q}}(\tau - \tau_1) G^{(0)}_{\mathbf{k}_2, \mathbf{k}_2-\mathbf{q}}(0) G^{(0)}_{\mathbf{k}_1, \mathbf{k}}(\tau_1) \\
&- G^{(0)}_{\mathbf{k}, \mathbf{k}_1+\mathbf{q}}(\tau - \tau_1) G^{(0)}_{\mathbf{k}_1, \mathbf{k}_2-\mathbf{q}}(\tau_1 - \tau_2) G^{(0)}_{\mathbf{k}_2, \mathbf{k}}(\tau_2) \Big]
\end{aligned} \quad \text{(A.4)}
$$

A.1 Electron Green's Function

Each term in the integral will be analyzed separately. Starting from the beginning and omitting both the sum and area prefactor for clarity I have

$$
\begin{aligned}
(1) &= \int_0^\beta d\tau\, e^{i\omega_n \tau} \int_0^\beta d\tau_1 \int_0^\beta d\tau_2\, \mathscr{D}(\mathbf{q}, \tau_1 - \tau_2)\, G^{(0)}_{\mathbf{k}_2, \mathbf{k}_1 + \mathbf{q}}(\tau_2 - \tau_1) \\
&\quad \times G^{(0)}_{\mathbf{k}_1, \mathbf{k}_2 - \mathbf{q}}(\tau_1 - \tau_2)\, G^{(0)}_{\mathbf{k}}(\tau) \\
&= G^{(0)}_{\mathbf{k}}(i\omega_n) \int_0^\beta d\tau_1 \int_0^\beta d\tau_2\, \mathscr{D}(\mathbf{q}, \tau_1 - \tau_2)\, G^{(0)}_{\mathbf{k}_2, \mathbf{k}_1 + \mathbf{q}}(\tau_2 - \tau_1)\, G^{(0)}_{\mathbf{k}_1, \mathbf{k}_2 - \mathbf{q}}(\tau_1 - \tau_2)
\end{aligned}
\tag{A.5}
$$

At this point the propagators remaining inside the integral are expressed using their Matsubara frequency Fourier transforms. Notice also that momentum conservation requires $\mathbf{k}_2 = \mathbf{k}_1 + \mathbf{q}$

$$
\begin{aligned}
(1) &= \frac{1}{\beta^3} \sum_{\mu, l, m} \mathscr{D}(\mathbf{q}, i\Omega_\mu)\, G^{(0)}_{\mathbf{k}}(i\omega_n)\, G^{(0)}_{\mathbf{k}_1 + \mathbf{q}}(i\omega_l)\, G^{(0)}_{\mathbf{k}_1}(i\omega_m) \\
&\quad \times \int_0^\beta d\tau_1 \int_0^\beta d\tau_2\, e^{-i(\Omega_\mu + \omega_m - \omega_l)(\tau_1 - \tau_2)}
\end{aligned}
\tag{A.6}
$$

The oscillating phase in the integral collapses the sum so that only the term with $\omega_l = \Omega_\mu + \omega_m$ survives. I also use the identities

$$
G^{(0)}_{\mathbf{k}}(i\omega_n) = \frac{1}{i\omega_n - \varepsilon_{\mathbf{k}}} \qquad \mathscr{D}(\mathbf{q}, i\Omega_\mu) = \frac{-2\omega_{\mathbf{q}}}{\Omega_\mu^2 + \omega_{\mathbf{q}}^2}
$$

and arrive at

$$
(1) = \frac{1}{\beta} \frac{1}{i\omega_n - \varepsilon_{\mathbf{k}}} \sum_{\mu, m} \frac{-2\omega_{\mathbf{q}}}{\Omega_\mu^2 + \omega_{\mathbf{q}}^2} \frac{1}{i\omega_m + i\Omega_\mu - \varepsilon_{\mathbf{k}_1 + \mathbf{q}}} \frac{1}{i\omega_m - \varepsilon_{\mathbf{k}_1}}
\tag{A.7}
$$

The second term in the integral has the form

$$
\begin{aligned}
(2) &= -\int_0^\beta d\tau\, e^{i\omega_n \tau} \int_0^\beta d\tau_1 \int_0^\beta d\tau_2\, \mathscr{D}(\mathbf{q}, \tau_1 - \tau_2) G^{(0)}_{\mathbf{k}_1, \mathbf{k}_1 + \mathbf{q}}(0)\, G^{(0)}_{\mathbf{k}_2, \mathbf{k}_2 - \mathbf{q}}(0)\, G^{(0)}_{\mathbf{k}}(\tau) \\
&= -G^{(0)}_{\mathbf{k}}(i\omega_n) \int_0^\beta d\tau_1 \int_0^\beta d\tau_2\, \mathscr{D}(\mathbf{q}, \tau_1 - \tau_2) G^{(0)}_{\mathbf{k}_1, \mathbf{k}_1 + \mathbf{q}}(0)\, G^{(0)}_{\mathbf{k}_2, \mathbf{k}_2 - \mathbf{q}}(0)
\end{aligned}
\tag{A.8}
$$

Proceeding in the same manner as before yields

$$
\begin{aligned}
(2) &= -\frac{1}{\beta^3} \sum_{\mu,l,m} \mathscr{D}(\mathbf{0}, i\Omega_\mu) \, G_\mathbf{k}^{(0)}(i\omega_n) \, G_{\mathbf{k}_1}^{(0)}(i\omega_l) \, G_{\mathbf{k}_2}^{(0)}(i\omega_m) \\
&\quad \times \int_0^\beta d\tau_1 \int_0^\beta d\tau_2 \, e^{-i\Omega_\mu(\tau_1 - \tau_2)} \\
&= -\frac{1}{\beta} \sum_{l,m} \mathscr{D}(\mathbf{0}, i0) \, G_\mathbf{k}^{(0)}(i\omega_n) \, G_{\mathbf{k}_1}^{(0)}(i\omega_l) \, G_{\mathbf{k}_2}^{(0)}(i\omega_m) \\
&= \frac{1}{\beta} \frac{2}{\omega_0} \frac{1}{i\omega_n - \varepsilon_\mathbf{k}} \sum_l \frac{1}{i\omega_l - \varepsilon_{\mathbf{k}_1}} \sum_m \frac{1}{i\omega_m - \varepsilon_{\mathbf{k}_2}} \\
&= \beta \frac{2}{\omega_0} \frac{1}{i\omega_n - \varepsilon_\mathbf{k}} \left(n_F(\varepsilon_{\mathbf{k}_1}) - \frac{1}{2} \right) \left(n_F(\varepsilon_{\mathbf{k}_2}) - \frac{1}{2} \right) \quad\quad (A.9)
\end{aligned}
$$

Moving on to the third term I have

$$
\begin{aligned}
(3) &= \int_0^\beta d\tau \, e^{i\omega_n \tau} \int_0^\beta d\tau_1 \int_0^\beta d\tau_2 \, \mathscr{D}(\mathbf{q}, \tau_1 - \tau_2) G_{\mathbf{k}_1,\mathbf{k}_1+\mathbf{q}}^{(0)}(0) \\
&\quad \times G_{\mathbf{k},\mathbf{k}_2-\mathbf{q}}^{(0)}(\tau - \tau_2) \, G_{\mathbf{k}_2,\mathbf{k}}^{(0)}(\tau_2) \\
&= \frac{1}{\beta^4} \sum_{\mu,l,m,j} \mathscr{D}(\mathbf{0}, i\Omega_\mu) \, G_{\mathbf{k}_1}^{(0)}(i\omega_l) \, G_\mathbf{k}^{(0)}(i\omega_m) \, G_\mathbf{k}^{(0)}(i\omega_j) \\
&\quad \times \int_0^\beta d\tau \, e^{-i(\omega_m - \omega_n)\tau} \int_0^\beta d\tau_1 \, e^{-i\Omega_\mu \tau_1} \int_0^\beta d\tau_2 \, e^{i(\Omega_\mu + \omega_m - \omega_j)\tau_2} \\
&= \frac{1}{\beta} \sum_l \mathscr{D}(\mathbf{0}, i0) \, G_{\mathbf{k}_1}^{(0)}(i\omega_l) \, G_\mathbf{k}^{(0)}(i\omega_n) \, G_\mathbf{k}^{(0)}(i\omega_n) \\
&= -\frac{2}{\omega_0} \left(\frac{1}{i\omega_n - \varepsilon_\mathbf{k}} \right)^2 \left(n_F(\varepsilon_{\mathbf{k}_1}) - \frac{1}{2} \right) \quad\quad (A.10)
\end{aligned}
$$

Continuing for the fourth I have

$$
\begin{aligned}
(4) &= -\int_0^\beta d\tau \, e^{i\omega_n \tau} \int_0^\beta d\tau_1 \int_0^\beta d\tau_2 \, \mathscr{D}(\mathbf{q}, \tau_1 - \tau_2) G_{\mathbf{k}_2,\mathbf{k}_1+\mathbf{q}}^{(0)}(\tau_2 - \tau_1) \\
&\quad \times G_{\mathbf{k},\mathbf{k}_2-\mathbf{q}}^{(0)}(\tau - \tau_2) \, G_{\mathbf{k}_1,\mathbf{k}}^{(0)}(\tau_1) \\
&= -\frac{1}{\beta^4} \sum_{\mu,l,m,j} \mathscr{D}(\mathbf{q}, i\Omega_\mu) \, G_{\mathbf{k}+\mathbf{q}}^{(0)}(i\omega_l) \, G_\mathbf{k}^{(0)}(i\omega_m) \, G_\mathbf{k}^{(0)}(i\omega_j)
\end{aligned}
$$

A.1 Electron Green's Function

$$\times \int_0^\beta d\tau\, e^{i(\omega_n - \omega_m)\tau} \int_0^\beta d\tau_1\, e^{-i(\Omega_\mu + \omega_j - \omega_l)\tau_1} \int_0^\beta d\tau_2\, e^{i(\Omega_\mu - \omega_l + \omega_m)\tau_2}$$

$$= -\frac{1}{\beta} \sum_\mu \mathscr{D}(\mathbf{q}, i\Omega_\mu)\, G^{(0)}_{\mathbf{k+q}}(i\omega_n + i\Omega_\mu)\, G^{(0)}_{\mathbf{k}}(i\omega_n)\, G^{(0)}_{\mathbf{k}}(i\omega_n)$$

$$= \frac{1}{\beta}\left(\frac{1}{i\omega_n - \varepsilon_{\mathbf{k}}}\right)^2 \sum_\mu \frac{2\omega_{\mathbf{q}}}{\Omega_\mu^2 + \omega_{\mathbf{q}}^2}\, \frac{1}{i\omega_n + i\Omega_\mu - \varepsilon_{\mathbf{k+q}}} \tag{A.11}$$

Evaluating the fifth yields

$$(5) = \int_0^\beta d\tau\, e^{i\omega_n \tau} \int_0^\beta d\tau_1 \int_0^\beta d\tau_2\, \mathscr{D}(\mathbf{q}, \tau_1 - \tau_2) G^{(0)}_{\mathbf{k},\mathbf{k}_1+\mathbf{q}}(\tau - \tau_1)$$

$$\times G^{(0)}_{\mathbf{k}_2,\mathbf{k}_2-\mathbf{q}}(0)\, G^{(0)}_{\mathbf{k}_1,\mathbf{k}}(\tau_1)$$

$$= \frac{1}{\beta^4} \sum_{\mu,l,m,j} \mathscr{D}(\mathbf{0}, i\Omega_\mu)\, G^{(0)}_{\mathbf{k}}(i\omega_l)\, G^{(0)}_{\mathbf{k}_2}(i\omega_m)\, G^{(0)}_{\mathbf{k}}(i\omega_j)$$

$$\times \int_0^\beta d\tau\, e^{i(\omega_n - \omega_l)\tau} \int_0^\beta d\tau_1\, e^{-i(\Omega_\mu - \omega_l + \omega_j)\tau_1} \int_0^\beta d\tau_2\, e^{i\Omega_\mu \tau_2}$$

$$= \frac{1}{\beta} \sum_m \mathscr{D}(\mathbf{0}, i0)\, G^{(0)}_{\mathbf{k}}(i\omega_n)\, G^{(0)}_{\mathbf{k}_2}(i\omega_m)\, G^{(0)}_{\mathbf{k}}(i\omega_n)$$

$$= -\frac{2}{\omega_0}\left(\frac{1}{i\omega_n - \varepsilon_{\mathbf{k}}}\right)^2 \left(n_F(\varepsilon_{\mathbf{k}_2}) - \frac{1}{2}\right) \tag{A.12}$$

And finally for the sixth I have

$$(6) = -\int_0^\beta d\tau\, e^{i\omega_n \tau} \int_0^\beta d\tau_1 \int_0^\beta d\tau_2\, \mathscr{D}(\mathbf{q}, \tau_1 - \tau_2) G^{(0)}_{\mathbf{k},\mathbf{k}_1+\mathbf{q}}(\tau - \tau_1)$$

$$\times G^{(0)}_{\mathbf{k}_1,\mathbf{k}_2-\mathbf{q}}(\tau_1 - \tau_2)\, G^{(0)}_{\mathbf{k}_2,\mathbf{k}}(\tau_2)$$

$$= -\frac{1}{\beta^4} \sum_{\mu,l,m,j} \mathscr{D}(\mathbf{q}, i\Omega_\mu)\, G^{(0)}_{\mathbf{k}}(i\omega_l)\, G^{(0)}_{\mathbf{k-q}}(i\omega_m)\, G^{(0)}_{\mathbf{k}}(i\omega_j)$$

$$\times \int_0^\beta d\tau\, e^{i(\omega_n - \omega_l)\tau} \int_0^\beta d\tau_1\, e^{i(\omega_l - \omega_m - \Omega_\mu)\tau_1} \int_0^\beta d\tau_2\, e^{i(\Omega_\mu + \omega_m - \omega_j)\tau_2}$$

$$= -\frac{1}{\beta} \sum_\mu \mathscr{D}(\mathbf{q}, i\Omega_\mu)\, G^{(0)}_{\mathbf{k}}(i\omega_n)\, G^{(0)}_{\mathbf{k-q}}(i\omega_n - i\Omega_\mu)\, G^{(0)}_{\mathbf{k}}(i\omega_n)$$

$$= \frac{1}{\beta}\left(\frac{1}{i\omega_n - \varepsilon_{\mathbf{k}}}\right)^2 \sum_\mu \frac{2\omega_{\mathbf{q}}}{\Omega_\mu^2 + \omega_{\mathbf{q}}^2}\, \frac{1}{i\omega_n - i\Omega_\mu - \varepsilon_{\mathbf{k-q}}} \tag{A.13}$$

Plugging all these results back into the expression for the dressed electron propagator leaves one with

$$
\begin{aligned}
G(\mathbf{k}, i\omega_n, T) = \frac{1}{\mathcal{A}} \frac{1}{i\omega_n - \varepsilon_\mathbf{k}} &\Bigg\{ \sum_{\mathbf{k}_1} |g_0|^2 \frac{-2}{\omega_0} \left(\frac{1}{i\omega_n - \varepsilon_\mathbf{k}} \right) \left(n_F(\varepsilon_{\mathbf{k}_1}) - \frac{1}{2} \right) \\
&+ \sum_{\mathbf{k}_2} |g_0|^2 \frac{-2}{\omega_0} \left(\frac{1}{i\omega_n - \varepsilon_\mathbf{k}} \right) \left(n_F(\varepsilon_{\mathbf{k}_2}) - \frac{1}{2} \right) \\
&+ \sum_{\mathbf{k}_1, \mathbf{k}_2} |g_0|^2 \frac{2\beta}{\omega_0} \left(n_F(\varepsilon_{\mathbf{k}_1}) - \frac{1}{2} \right) \left(n_F(\varepsilon_{\mathbf{k}_2}) - \frac{1}{2} \right) \\
&+ \sum_{\mathbf{k}_1, \mathbf{q}} |g_\mathbf{q}|^2 \frac{1}{\beta} \sum_m \frac{1}{i\omega_m - \varepsilon_{\mathbf{k}_1}} \sum_\mu \frac{-2\omega_\mathbf{q}}{\Omega_\mu^2 + \omega_\mathbf{q}^2} \frac{1}{i\omega_m + i\Omega_\mu - \varepsilon_{\mathbf{k}_1 + \mathbf{q}}} \\
&+ \sum_\mathbf{q} |g_\mathbf{q}|^2 \frac{1}{\beta} \frac{1}{i\omega_n - \varepsilon_\mathbf{k}} \sum_\mu \frac{2\omega_\mathbf{q}}{\Omega_\mu^2 + \omega_\mathbf{q}^2} \\
&\times \left(\frac{1}{i\omega_n + i\Omega_\mu - \varepsilon_{\mathbf{k}+\mathbf{q}}} + \frac{1}{i\omega_n - i\Omega_\mu - \varepsilon_{\mathbf{k}-\mathbf{q}}} \right) \Bigg\}
\end{aligned}
\tag{A.14}
$$

To simplify this expression I first perform the sums over the bosonic frequencies (that is the sums over μ) as described in (A.2). Doing so yields

$$
\begin{aligned}
G(\mathbf{k}, i\omega_n, T) = \frac{1}{\mathcal{A}} \frac{1}{i\omega_n - \varepsilon_\mathbf{k}} &\Bigg\{ \sum_{\mathbf{k}_1} |g_0|^2 \frac{-2}{\omega_0} \left(\frac{1}{i\omega_n - \varepsilon_\mathbf{k}} \right) \left(n_F(\varepsilon_{\mathbf{k}_1}) - \frac{1}{2} \right) \\
&+ \sum_{\mathbf{k}_2} |g_0|^2 \frac{-2}{\omega_0} \left(\frac{1}{i\omega_n - \varepsilon_\mathbf{k}} \right) \left(n_F(\varepsilon_{\mathbf{k}_2}) - \frac{1}{2} \right) \\
&+ \sum_{\mathbf{k}_1, \mathbf{k}_2} |g_0|^2 \frac{2\beta}{\omega_0} \left(n_F(\varepsilon_{\mathbf{k}_1}) - \frac{1}{2} \right) \left(n_F(\varepsilon_{\mathbf{k}_2}) - \frac{1}{2} \right) \\
&+ \sum_{\mathbf{k}_1, \mathbf{q}} |g_\mathbf{q}|^2 \sum_m \frac{-1}{i\omega_m - \varepsilon_{\mathbf{k}_1}} \Bigg[\frac{n_B(\omega_\mathbf{q}) + n_F(\varepsilon_{\mathbf{k}_1+\mathbf{q}})}{i\omega_m - \varepsilon_{\mathbf{k}_1+\mathbf{q}} + \omega_\mathbf{q}} \\
&+ \frac{n_B(\omega_\mathbf{q}) + 1 - n_F(\varepsilon_{\mathbf{k}_1+\mathbf{q}})}{i\omega_m - \varepsilon_{\mathbf{k}_1+\mathbf{q}} - \omega_\mathbf{q}} \Bigg] + \sum_\mathbf{q} |g_\mathbf{q}|^2 \frac{1}{i\omega_n - \varepsilon_\mathbf{k}} \\
&\times \Bigg[\frac{n_B(\omega_\mathbf{q}) + n_F(\varepsilon_{\mathbf{k}+\mathbf{q}})}{i\omega_n - \varepsilon_{\mathbf{k}+\mathbf{q}} + \omega_\mathbf{q}} + \frac{n_B(\omega_\mathbf{q}) + 1 - n_F(\varepsilon_{\mathbf{k}+\mathbf{q}})}{i\omega_n - \varepsilon_{\mathbf{k}+\mathbf{q}} - \omega_\mathbf{q}} \\
&+ \frac{n_B(\omega_\mathbf{q}) + n_F(\varepsilon_{\mathbf{k}-\mathbf{q}})}{i\omega_n - \varepsilon_{\mathbf{k}-\mathbf{q}} + \omega_\mathbf{q}} + \frac{n_B(\omega_\mathbf{q}) + 1 - n_F(\varepsilon_{\mathbf{k}-\mathbf{q}})}{i\omega_n - \varepsilon_{\mathbf{k}-\mathbf{q}} - \omega_\mathbf{q}} \Bigg] \Bigg\}
\end{aligned}
\tag{A.15}
$$

A.2 Bosonic Sums

At this point I could continue to evaluate the remaining fermionic sum over m. However, for my purpose here it is unnecessary. Note that the diagram for the DFQ self energy, much like the phonon self-energy, requires two external fermion lines with momentum $\mathbf{k}$ and energy ω (or rather $i\omega_n$ since I have not yet performed analytic continuation). The only term in (A.15) that satisfies this requirement and accounts for the creation and annihilation of phonons is the final sum. Taking into account the fact that the external fermion lines are not included in the expression for the self-energy $\Sigma(\mathbf{k}, i\omega_n, T)$ it can be immediately identified as

$$\Sigma(\mathbf{k}, i\omega_n, T) = \frac{1}{\mathcal{A}} \sum_{\mathbf{q}} |g_{\mathbf{q}}|^2 \left\{ \frac{1}{2}\left(1 + \hat{\mathbf{k}} \cdot \widehat{\mathbf{k}+\mathbf{q}}\right) \left[\frac{n_B(\omega_{\mathbf{q}}) + n_F(\varepsilon_{\mathbf{k}+\mathbf{q}})}{i\omega_n - \varepsilon_{\mathbf{k}+\mathbf{q}} + \omega_{\mathbf{q}}} \right. \right.$$
$$\left. + \frac{n_B(\omega_{\mathbf{q}}) + 1 - n_F(\varepsilon_{\mathbf{k}+\mathbf{q}})}{i\omega_n - \varepsilon_{\mathbf{k}+\mathbf{q}} - \omega_{\mathbf{q}}} \right] + \frac{1}{2}\left(1 + \hat{\mathbf{k}} \cdot \widehat{\mathbf{k}-\mathbf{q}}\right)$$
$$\left. \times \left[\frac{n_B(\omega_{\mathbf{q}}) + n_F(\varepsilon_{\mathbf{k}-\mathbf{q}})}{i\omega_n - \varepsilon_{\mathbf{k}-\mathbf{q}} + \omega_{\mathbf{q}}} + \frac{n_B(\omega_{\mathbf{q}}) + 1 - n_F(\varepsilon_{\mathbf{k}-\mathbf{q}})}{i\omega_n - \varepsilon_{\mathbf{k}-\mathbf{q}} - \omega_{\mathbf{q}}} \right] \right\} \quad (A.16)$$

where the factors $\frac{1}{2}\left(1 + \hat{\mathbf{k}} \cdot \widehat{\mathbf{k} \pm \mathbf{q}}\right)$ have been added by hand to account for the spin chirality of the DFQs at the Fermi surface. This expression can be simplified by realizing that, because the sum is over all $\mathbf{q}$, the $\mathbf{k}+\mathbf{q}$ and $\mathbf{k}-\mathbf{q}$ terms contribute equally. This leaves

$$\Sigma(\mathbf{k}, i\omega_n, T) = \frac{1}{\mathcal{A}} \sum_{\mathbf{q}} |g_{\mathbf{q}}|^2 \left(1 + \hat{\mathbf{k}} \cdot \widehat{\mathbf{k}+\mathbf{q}}\right) \left[\frac{n_B(\omega_{\mathbf{q}}) + n_F(\varepsilon_{\mathbf{k}+\mathbf{q}})}{i\omega_n - \varepsilon_{\mathbf{k}+\mathbf{q}} + \omega_{\mathbf{q}}} \right.$$
$$\left. + \frac{n_B(\omega_{\mathbf{q}}) + 1 - n_F(\varepsilon_{\mathbf{k}+\mathbf{q}})}{i\omega_n - \varepsilon_{\mathbf{k}+\mathbf{q}} - \omega_{\mathbf{q}}} \right] \quad (A.17)$$

which reproduces Eq. (7.2).

A.2 Bosonic Sums

The Bosonic sums of the previous section have the form

$$\sum_{\mu} \frac{-2\omega_{\mathbf{q}}}{\Omega_{\mu}^2 + \omega_{\mathbf{q}}^2} \frac{1}{i\omega_n \pm i\Omega_{\mu} - \varepsilon_{\mathbf{k}+\mathbf{q}}} \quad (A.18)$$

To evaluate them I replace the sum by a contour integration circling the imaginary axis using the g function with poles of residue 1 at the bosonic frequencies.

$$g(z) = \beta \frac{1}{e^{\beta z} - 1} \quad (A.19)$$

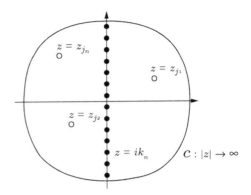

Fig. A.1 The analytic continuation procedure in the complex z-plane where the Matsubara function defined for $z = \omega_n$ goes to the retarded or advanced Green's functions defined infinitesimally close to real axis

with

$$S \equiv \sum_\mu h(\Omega_\mu) = \frac{1}{2\pi i} \oint dz\, g(z)\, h(-iz) \tag{A.20}$$

I then express the h function in the following manner

$$h(-iz) = \begin{cases} h_1 = \dfrac{-2\omega_q}{-z^2 + \omega_q^2} \dfrac{1}{i\omega_n + z - \varepsilon_{k+q}} \\ = \dfrac{2\omega_q}{(z+\omega_q)(z-\omega_q)(i\omega_n + z - \varepsilon_{k+q})} \\ h_2 = \dfrac{-2\omega_q}{-z^2 + \omega_q^2} \dfrac{1}{i\omega_n - z - \varepsilon_{k+q}} \\ = \dfrac{2\omega_q}{(z+\omega_q)(z-\omega_q)(i\omega_n - z - \varepsilon_{k+q})} \end{cases} \tag{A.21}$$

This function has three poles, two at $z = \pm\omega_q$ and one at $z = \pm(\varepsilon_{k+q} - i\omega_n)$ (Fig. A.1).

Deforming the contour to infinity while avoiding the crossing of these singularities leads one to the conclusion that the Matsubara summation is equal to -1 times the sum of the residues of the product gh at these three poles. The minus sign comes from the fact that the contour now circulates clockwise around the three poles. The residues are

$$R(z = \omega_q) = \beta\, n_B(\omega_q) \frac{2\omega_q}{\omega_q + \omega_q} \frac{1}{i\omega_n \pm \omega_q - \varepsilon_{k+q}}$$

$$= \frac{\beta\, n_B(\omega_q)}{i\omega_n \pm \omega_q - \varepsilon_{k+q}}$$

A.2 Bosonic Sums

$$R(z = -\omega_\mathbf{q}) = \beta\, n_B(-\omega_\mathbf{q})\, \frac{2\omega_\mathbf{q}}{-\omega_\mathbf{q} - \omega_\mathbf{q}}\, \frac{1}{i\omega_n \mp \omega_\mathbf{q} - \varepsilon_{\mathbf{k+q}}}$$

$$= \frac{-\beta\, n_B(-\omega_\mathbf{q})}{i\omega_n \mp \omega_\mathbf{q} - \varepsilon_{\mathbf{k+q}}}$$

$$R(z = \varepsilon_{\mathbf{k+q}} - i\omega_n) = \frac{2\omega_\mathbf{q}\, \beta}{(\varepsilon_{\mathbf{k+q}} - i\omega_n + \omega_\mathbf{q})(\varepsilon_{\mathbf{k+q}} - i\omega_n - \omega_\mathbf{q})\left(e^{\beta(\varepsilon_{\mathbf{k+q}} - i\omega_n)} - 1\right)}$$

$$R(z = i\omega_n - \varepsilon_{\mathbf{k+q}}) = \frac{-2\omega_\mathbf{q}\, \beta}{(i\omega_n - \varepsilon_{\mathbf{k+q}} + \omega_\mathbf{q})(i\omega_n - \varepsilon_{\mathbf{k+q}} - \omega_\mathbf{q})\left(e^{\beta(i\omega_n - \varepsilon_{\mathbf{k+q}})} - 1\right)}$$
(A.22)

Since $\exp[\pm i\beta\omega_n] = -1$ for fermionic frequencies ω_n

$$R(z = \varepsilon_{\mathbf{k+q}} - i\omega_n) = \frac{-2\omega_\mathbf{q}\, \beta}{(\varepsilon_{\mathbf{k+q}} - i\omega_n + \omega_\mathbf{q})(\varepsilon_{\mathbf{k+q}} - i\omega_n - \omega_\mathbf{q})}\, n_F(\varepsilon_{\mathbf{k+q}})$$

$$= \frac{\beta\, n_F(\varepsilon_{\mathbf{k+q}})}{i\omega_n + \omega_\mathbf{q} - \varepsilon_{\mathbf{k+q}}} - \frac{\beta\, n_F(\varepsilon_{\mathbf{k+q}})}{i\omega_n - \omega_\mathbf{q} - \varepsilon_{\mathbf{k+q}}}$$

$$R(z = i\omega_n - \varepsilon_{\mathbf{k+q}}) = \frac{2\omega_\mathbf{q}\, \beta}{(i\omega_n - \varepsilon_{\mathbf{k+q}} + \omega_\mathbf{q})(i\omega_n - \varepsilon_{\mathbf{k+q}} - \omega_\mathbf{q})}\, n_F(-\varepsilon_{\mathbf{k+q}})$$

$$= \frac{\beta\, n_F(-\varepsilon_{\mathbf{k+q}})}{i\omega_n - \omega_\mathbf{q} - \varepsilon_{\mathbf{k+q}}} - \frac{\beta\, n_F(-\varepsilon_{\mathbf{k+q}})}{i\omega_n + \omega_\mathbf{q} - \varepsilon_{\mathbf{k+q}}}$$
(A.23)

However, noting that

$$n_B(-x) = -1 - n_B(x), \quad \text{and} \quad n_F(-x) = 1 - n_F(x)$$
(A.24)

I obtain the same results for either form of the sum, namely

$$S_1 = S_2 = -\beta \left[\frac{n_B(\omega_\mathbf{q}) + n_F(\varepsilon_{\mathbf{k+q}})}{i\omega_n - \varepsilon_{\mathbf{k+q}} + \omega_\mathbf{q}} + \frac{n_B(\omega_\mathbf{q}) + 1 - n_F(\varepsilon_{\mathbf{k+q}})}{i\omega_n - \varepsilon_{\mathbf{k+q}} - \omega_\mathbf{q}} \right]$$
(A.25)

Appendix B
Numerical Evaluation of the DFQ Self-Energy

The analysis of the DFQ self-energy in Appendix A yields the complex function

$$\Sigma(\mathbf{k}, i\omega_n, T) = \frac{1}{\mathcal{A}} \sum_{\mathbf{q}} |g_{\mathbf{q}}|^2 \left(1 + \hat{\mathbf{k}} \cdot \widehat{\mathbf{k}+\mathbf{q}}\right)$$
$$\times \left[\frac{n_B(\omega_{\mathbf{q}}) + n_F(\varepsilon_{\mathbf{k}+\mathbf{q}})}{i\omega_n - \varepsilon_{\mathbf{k}+\mathbf{q}} + \omega_{\mathbf{q}}} + \frac{n_B(\omega_{\mathbf{q}}) + 1 - n_F(\varepsilon_{\mathbf{k}+\mathbf{q}})}{i\omega_n - \varepsilon_{\mathbf{k}+\mathbf{q}} - \omega_{\mathbf{q}}} \right] \quad \text{(B.1)}$$

where $\mathbf{k}$ is the DFQ wave vector, $\mathbf{q}$ is the phonon wavevector, $g_{\mathbf{q}}$ is the electron-phonon matrix element, $\omega_{\mathbf{q}}$ is the phonon energy, $i\omega_n$ is the Matsubara frequency of the DFQs, $\varepsilon_{\mathbf{k}}$ is the nominal dispersion of the DFQ states, and n_B and n_F are the Bose and Fermi distributions, respectively. Knowledge of both the real and imaginary parts of $\Sigma(\mathbf{k}, i\omega_n, T)$ is necessary to calculate the DFQ spectral function. This appendix details the manner in which the real and imaginary parts were evaluated computationally.

Performing analytic continuation ($i\omega_n \to \omega + i\eta$) allows one to write

$$\Sigma(\mathbf{k}, \omega, T) = \frac{1}{\mathcal{A}} \sum_{\mathbf{q}} |g_{\mathbf{q}}|^2 \left(1 + \hat{\mathbf{k}} \cdot \widehat{\mathbf{k}+\mathbf{q}}\right)$$
$$\times \left[\frac{n_B(\omega_{\mathbf{q}}) + n_F(\varepsilon_{\mathbf{k}+\mathbf{q}})}{\omega + i\eta - \varepsilon_{\mathbf{k}+\mathbf{q}} + \omega_{\mathbf{q}}} + \frac{n_B(\omega_{\mathbf{q}}) + 1 - n_F(\varepsilon_{\mathbf{k}+\mathbf{q}})}{\omega + i\eta - \varepsilon_{\mathbf{k}+\mathbf{q}} - \omega_{\mathbf{q}}} \right] \quad \text{(B.2)}$$

At this point I convert the sum to an integral by using the transformation

$$\sum_{\mathbf{q}} \to \frac{\mathcal{A}}{(2\pi)^2} \int d\mathbf{q} \quad \text{(B.3)}$$

and employ the principal value theorem

$$\frac{1}{a+\omega+i\eta} = \mathcal{P}\left(\frac{1}{a+\omega}\right) - i\pi\delta(a+\omega) \tag{B.4}$$

to arrive at the imaginary part of the expression

$$\mathrm{Im}[\Sigma(\mathbf{k},\omega,T)] = -\frac{1}{4\pi}\int dq\, q \int d\varphi\, |g_\mathbf{q}|^2 \left(1 + \hat{\mathbf{k}}\cdot\widehat{\mathbf{k}+\mathbf{q}}\right)$$
$$\times\left[\Big(n_B(\omega_\mathbf{q}) + n_F(\varepsilon_{\mathbf{k}+\mathbf{q}})\Big)\delta(\omega - \varepsilon_{\mathbf{k}+\mathbf{q}} + \omega_\mathbf{q})\right.$$
$$\left.+\Big(n_B(\omega_\mathbf{q}) + 1 - n_F(\varepsilon_{\mathbf{k}+\mathbf{q}})\Big)\delta(\omega - \varepsilon_{\mathbf{k}+\mathbf{q}} - \omega_\mathbf{q})\right] \tag{B.5}$$

B.1 Hole Term

B.1.1 Above Dirac Point

I will begin by concentrating on the first term in the expression above. From this point on I will make the small but important substitution $\omega \to \hbar\omega$ and $\omega_\mathbf{q} \to \hbar\omega_\mathbf{q}$. Thus ω and $\omega_\mathbf{q}$ now have the conventional unit of frequency. Also, recall that above the Dirac point one can write the DFQ dispersion as $\varepsilon_{\mathbf{k}+\mathbf{q}} = \hbar v_0(|\mathbf{k}+\mathbf{q}| - k_F)$. Note that the delta function requires its argument to be zero in order to yield finite results. Thus I am interested in the case

$$f(\cos\varphi_0) = \hbar\omega - \hbar v_0\sqrt{k^2 + q^2 + 2kq\cos\varphi_0} + \hbar v_0 k_F + \hbar\omega_\mathbf{q} = 0 \tag{B.6}$$

Defining $z \equiv \cos\varphi$, $\Omega \equiv \omega/v_0$, and $\Omega_\mathbf{q} \equiv \omega_\mathbf{q}/v_0$ I can express this condition as

$$z_0 = \frac{(\Omega + \Omega_\mathbf{q} + k_F)^2 - k^2 - q^2}{2kq} \tag{B.7}$$

where the subscript on z indicates that it is the value that satisfies the delta function. Since z is bounded by -1 and 1, I am led to the restrictions

$$(k-q)^2 \leq (\Omega + \Omega_\mathbf{q} + k_F)^2 \qquad (\Omega + \Omega_\mathbf{q} + k_F)^2 \leq (k+q)^2 \tag{B.8}$$

Taking the square roots and being careful with the signs yields four possibilities for each inequality

$$(k-q) \leq (\Omega + \Omega_\mathbf{q} + k_F) \qquad (\Omega + \Omega_\mathbf{q} + k_F) \leq (k+q)$$
$$(k-q) \geq -(\Omega + \Omega_\mathbf{q} + k_F) \qquad (\Omega + \Omega_\mathbf{q} + k_F) \geq -(k+q)$$

B.1 Hole Term

$$-(k-q) \leq (\Omega + \Omega_q + k_F) \qquad -(\Omega + \Omega_q + k_F) \leq (k+q)$$
$$-(k-q) \geq -(\Omega + \Omega_q + k_F) \qquad -(\Omega + \Omega_q + k_F) \geq -(k+q) \tag{B.9}$$

In each column there are four inequalities but only two are unique. Between the two columns, then, I arrive at a total of four inequalities that I present below in terms of Heaviside-Theta functions.

$$\Theta(\Omega + \Omega_q + k_F - k + q)$$
$$\Theta(k - q + \Omega + \Omega_q + k_F)$$
$$\Theta(k + q - \Omega - \Omega_q - k_F)$$
$$\Theta(\Omega + \Omega_q + k_F + k + q) \tag{B.10}$$

If these four conditions are not met, then the delta function is not satisfied, so I write

$$\delta(f(z)) = \frac{\delta(z - z_0)}{|f'(z_0)|} \Theta(\Omega + \Omega_q + k_F - k + q)\Theta(k - q + \Omega + \Omega_q + k_F)$$
$$\times \Theta(k + q - \Omega - \Omega_q - k_F)\Theta(\Omega + \Omega_q + k_F + k + q) \tag{B.11}$$

Now I evaluate the derivative $f'(z)$

$$f'(z) = -\frac{\hbar v_0 kq}{\sqrt{k^2 + q^2 + 2kqz}} \tag{B.12}$$

and find

$$f'(z_0) = -\frac{\hbar v_0 kq}{\Omega + \Omega_q + k_F} \tag{B.13}$$

I can now rewrite the delta function in the more convenient variable z as

$$\delta(f(z)) = \delta(z - z_0) \frac{\Omega + \Omega_q + k_F}{\hbar v_0 kq} \Theta(\Omega + \Omega_q + k_F - k + q)\Theta(k - q + \Omega + \Omega_q + k_F)$$
$$\times \Theta(k + q - \Omega - \Omega_q - k_F)\Theta(\Omega + \Omega_q + k_F + k + q) \tag{B.14}$$

Having simplified the delta function, I now turn to evaluating the chirality factor.

$$(1 + \hat{\mathbf{k}} \cdot \widehat{\mathbf{k} + \mathbf{q}}) = \left(1 + \frac{k + qz}{\sqrt{k^2 + q^2 + 2kqz}}\right) \tag{B.15}$$

Its value at z_0 can be found by performing the necessary algebra.

$$\left(1 + \frac{k + qz_0}{\sqrt{k^2 + q^2 + 2kqz_0}}\right) = \left(1 + \frac{k + qz_0}{\Omega + \Omega_0 + k_F}\right)$$

$$= \frac{1}{2k} \frac{1}{\Omega + \Omega_0 + k_F} \left(2k(\Omega + \Omega_0 + k_F) + 2k^2\right.$$

$$\left. + (\Omega + \Omega_0 + k_F)^2 - k^2 - q^2\right)$$

$$= \frac{(k + \Omega + \Omega_0 + k_F)^2 - q^2}{2k(\Omega + \Omega_0 + k_F)}$$

$$= \frac{(k + \Omega + \Omega_0 + k_F + q)(k + \Omega + \Omega_0 + k_F - q)}{2k(\Omega + \Omega_0 + k_F)}$$

(B.16)

Now I can deal with the Jacobian of the transformation of variables $\varphi \to z$

$$dz = -\sin\varphi \, d\varphi$$

$$d\varphi = -\frac{dz}{\sin\varphi} \quad (B.17)$$

keeping in mind that

$$\sin\varphi = \begin{cases} \sqrt{1 - \cos^2\varphi} = \sqrt{1 - z^2}, & 0 \le \varphi \le \pi \\ -\sqrt{1 - \cos^2\varphi} = -\sqrt{1 - z^2}, & \pi \le \varphi \le 2\pi \end{cases} \quad (B.18)$$

At this point I am in a position to rewrite the integral for the first term in Eq. (B.5) in the variable z. Specifically, I have

$$-\frac{1}{4\pi} \int dq \, q \, |g_q|^2 \left(\int_1^{-1} \frac{-dz}{\sqrt{1-z^2}} h(z)\delta(z - z_0) + \int_{-1}^1 \frac{dz}{\sqrt{1-z^2}} h(z)\delta(z - z_0)\right)$$

$$= -\frac{1}{2\pi} \int dq \, q \, |g_q|^2 \int_{-1}^1 \frac{dz}{\sqrt{1-z^2}} h(z)\delta(z - z_0) \quad (B.19)$$

with $h(z)$ defined as

$$h(z) = \left(1 + \frac{k + qz}{\sqrt{k^2 + q^2 + 2kqz}}\right) \frac{\Omega + \Omega_q + k_F}{\hbar v_0 kq}$$

$$\times \left(\frac{1}{e^{\beta\hbar v_0 \Omega_q} - 1} + \frac{1}{e^{\beta\hbar v_0(\sqrt{k^2 + q^2 + 2kqz} - k_F)} + 1}\right)$$

$$\times \Theta(\Omega + \Omega_q + k_F - k + q)\Theta(k - q + \Omega + \Omega_q + k_F)$$

$$\times \Theta(k + q - \Omega - \Omega_q - k_F)\Theta(\Omega + \Omega_q + k_F + k + q) \quad (B.20)$$

B.1 Hole Term

The delta function collapses the integral, leaving

$$-\frac{1}{2\pi}\int dq\, q\, |g_q|^2 \frac{h(z_0)}{\sqrt{1-z_0^2}} \tag{B.21}$$

Evaluating $h(z_0)$ and $1/\sqrt{1-z_0^2}$ yields

$$\begin{aligned}h(z_0) =& \frac{(k+\Omega+\Omega_\mathbf{q}+k_F+q)(k+\Omega+\Omega_\mathbf{q}+k_F-q)}{2k(\Omega+\Omega_\mathbf{q}+k_F)} \frac{\Omega+\Omega_\mathbf{q}+k_F}{\hbar v_0 k q} \\ & \times \left(\frac{1}{e^{\beta \hbar v_0 \Omega_q}-1} + \frac{1}{e^{\beta \hbar v_0 (\Omega+\Omega_\mathbf{q})}+1} \right) \\ & \times \Theta(\Omega+\Omega_\mathbf{q}+k_F-k+q)\Theta(k-q+\Omega+\Omega_\mathbf{q}+k_F) \\ & \times \Theta(k+q-\Omega-\Omega_\mathbf{q}-k_F)\Theta(\Omega+\Omega_\mathbf{q}+k_F+k+q) \end{aligned} \tag{B.22}$$

$$\begin{aligned}(1-z_0^2)^{-1/2} &= [(1-z_0)(1+z_0)]^{-1/2} \\ &= \frac{2kq}{\sqrt{(q-k+\Omega+\Omega_\mathbf{q}+k_F)(k+q-\Omega-\Omega_\mathbf{q}-k_F)(k-q+\Omega+\Omega_\mathbf{q}+k_F)(k+q+\Omega+\Omega_\mathbf{q}+k_F)}}\end{aligned} \tag{B.23}$$

Plugging everything in I arrive at

$$\begin{aligned}& -\frac{1}{2\pi \hbar v_0 k} \int dq\, q\, |g_q|^2 \\ & \times \frac{(k+\Omega+\Omega_\mathbf{q}+k_F+q)(k+\Omega+\Omega_\mathbf{q}+k_F-q)}{\sqrt{(q-k+\Omega+\Omega_\mathbf{q}+k_F)(k+q-\Omega-\Omega_\mathbf{q}-k_F)(k-q+\Omega+\Omega_\mathbf{q}+k_F)(k+q+\Omega+\Omega_\mathbf{q}+k_F)}} \\ & \times \left(\frac{1}{e^{\beta \hbar v_0 \Omega_q}-1} + \frac{1}{e^{\beta \hbar v_0 (\Omega+\Omega_\mathbf{q})}+1} \right) \\ & \times \Theta(\Omega+\Omega_\mathbf{q}+k_F-k+q)\Theta(k-q+\Omega+\Omega_\mathbf{q}+k_F) \\ & \times \Theta(k+q-\Omega-\Omega_\mathbf{q}-k_F)\Theta(\Omega+\Omega_\mathbf{q}+k_F+k+q)\end{aligned} \tag{B.24}$$

which simplifies to

$$\begin{aligned}& -\frac{1}{2\pi \hbar v_0 k} \int_0^{2k_F} dq\, q\, |g_q|^2 \sqrt{\frac{(k+\Omega+\Omega_\mathbf{q}+k_F+q)(k+\Omega+\Omega_\mathbf{q}+k_F-q)}{(q-k+\Omega+\Omega_\mathbf{q}+k_F)(k+q-\Omega-\Omega_\mathbf{q}-k_F)}} \\ & \times \left(\frac{1}{e^{\beta \hbar v_0 \Omega_q}-1} + \frac{1}{e^{\beta \hbar v_0 (\Omega+\Omega_\mathbf{q})}+1} \right) \\ & \times \Theta(\Omega+\Omega_\mathbf{q}+k_F-k+q)\Theta(k-q+\Omega+\Omega_\mathbf{q}+k_F) \\ & \times \Theta(k+q-\Omega-\Omega_\mathbf{q}-k_F)\Theta(\Omega+\Omega_\mathbf{q}+k_F+k+q)\end{aligned} \tag{B.25}$$

One will notice that the theta functions require each term in parentheses inside the radical to be positive, guaranteeing a real value for the integral. The fourth theta function is somewhat superfluous since, in this regime, $\Omega > -k_F$ and all the other terms are positive quantities.

B.1.2 Below Dirac Point

In the previous analysis I used the dispersion $\varepsilon_{\mathbf{k+q}} = \hbar v_0(|\mathbf{k} + \mathbf{q}| - k_F)$. However, this is only strictly correct for states above the Dirac point. For DFQ states below the Dirac point one needs to use a different expression. This subtlety comes from the fact that $|\mathbf{k} + \mathbf{q}|$, being the magnitude of a vector, is inherently positive. Thus, for states below the Dirac point I instead use $\varepsilon_{\mathbf{k+q}} = \hbar v_0(-|\mathbf{k} + \mathbf{q}| - k_F)$. One should also notice that in this regime $\Omega \leq -k_F$. Taking this into account requires nothing more than carefully repeating the previously outlined steps with the modified dispersion and keeping track of minus signs. The end result is that the integral expression in Eq. (B.25) need only be modified by changing the sign on Ω, $\Omega_{\mathbf{q}}$, and k_F for all instances *inside* the radical and theta functions (not in the Bose and Fermi factors). The result is

$$-\frac{1}{2\pi\hbar v_0 k} \int_0^{2k_F} dq\, q\, |g_q|^2 \sqrt{\frac{(k - \Omega - \Omega_{\mathbf{q}} - k_F + q)(k - \Omega - \Omega_{\mathbf{q}} - k_F - q)}{(q - k - \Omega - \Omega_{\mathbf{q}} - k_F)(k + q + \Omega + \Omega_{\mathbf{q}} + k_F)}}$$

$$\times \left(\frac{1}{e^{\beta \hbar v_0 \Omega_q} - 1} + \frac{1}{e^{\beta \hbar v_0 (\Omega + \Omega_q)} + 1} \right)$$

$$\times \Theta(-\Omega - \Omega_{\mathbf{q}} - k_F - k + q) \Theta(k - q - \Omega - \Omega_{\mathbf{q}} - k_F)$$

$$\times \Theta(k + q + \Omega + \Omega_{\mathbf{q}} + k_F) \Theta(-\Omega - \Omega_{\mathbf{q}} - k_F + k + q) \quad (B.26)$$

B.2 Particle Term

B.2.1 Above Dirac Point

Recall that the integrals in Eqs. (B.25) and (B.26) only take care of the first term in Eq. (B.5). I still need to evaluate the second. The two major differences are the change $n_F(\varepsilon_{\mathbf{k+q}}) \to 1 - n_F(\varepsilon_{\mathbf{k+q}})$ and the change of sign on ω_q in the delta function. I'll start by analyzing the modification to Fermi occupation term. First, I write

$$1 - n_F(\varepsilon_{\mathbf{k+q}}) = n_F(-\varepsilon_{\mathbf{k+q}}) = \frac{1}{e^{\hbar v_0 (k_F - \sqrt{k^2 + q^2 + 2kqz})} + 1} \quad (B.27)$$

B.2 Particle Term

Now notice that the value of z that satisfies the delta function changes slightly because of the change in sign of $\Omega_{\mathbf{q}}$. Specifically,

$$z'_0 = \frac{(\Omega - \Omega_{\mathbf{q}} + k_F)^2 - k^2 - q^2}{2kq} \qquad (B.28)$$

Thus, after the z-integration one can express the Fermi factor as

$$\frac{1}{e^{\hbar v_0(k_F - \sqrt{k^2 + q^2 + 2kqz'_0})} + 1} = \frac{1}{e^{-\Omega + \Omega_{\mathbf{q}}} + 1} \qquad (B.29)$$

The rest of the analysis proceeds exactly as before, with only a change of sign in $\Omega_{\mathbf{q}}$. The final integral expression is

$$-\frac{1}{2\pi \hbar v_0 k} \int_0^{2k_F} dq\, q\, |g_q|^2 \sqrt{\frac{(k + \Omega - \Omega_{\mathbf{q}} + k_F + q)(k + \Omega - \Omega_{\mathbf{q}} + k_F - q)}{(q - k + \Omega - \Omega_{\mathbf{q}} + k_F)(k + q - \Omega + \Omega_{\mathbf{q}} - k_F)}}$$

$$\times \left(\frac{1}{e^{\beta \hbar v_0 \Omega_q} - 1} + \frac{1}{e^{\beta \hbar v_0(-\Omega + \Omega_q)} + 1} \right)$$

$$\times \Theta(\Omega - \Omega_{\mathbf{q}} + k_F - k + q)\Theta(k - q + \Omega - \Omega_{\mathbf{q}} + k_F)$$

$$\times \Theta(k + q - \Omega + \Omega_{\mathbf{q}} - k_F)\Theta(\Omega - \Omega_{\mathbf{q}} + k_F + k + q) \qquad (B.30)$$

B.2.2 Below Dirac Point

The particle contribution from below the Dirac point can be obtained by making the exact same transformation to Eq. (B.30) that was made for the holes. For the sake of brevity I just state the result below.

$$-\frac{1}{2\pi \hbar v_0 k} \int_0^{2k_F} dq\, q\, |g_q|^2 \sqrt{\frac{(k - \Omega + \Omega_{\mathbf{q}} - k_F + q)(k - \Omega + \Omega_{\mathbf{q}} - k_F - q)}{(q - k - \Omega + \Omega_{\mathbf{q}} - k_F)(k + q + \Omega - \Omega_{\mathbf{q}} + k_F)}}$$

$$\times \left(\frac{1}{e^{\beta \hbar v_0 \Omega_q} - 1} + \frac{1}{e^{\beta \hbar v_0(-\Omega + \Omega_q)} + 1} \right)$$

$$\times \Theta(-\Omega + \Omega_{\mathbf{q}} - k_F - k + q)\Theta(k - q - \Omega + \Omega_{\mathbf{q}} - k_F)$$

$$\times \Theta(k + q + \Omega - \Omega_{\mathbf{q}} + k_F)\Theta(-\Omega + \Omega_{\mathbf{q}} - k_F + k + q) \qquad (B.31)$$

B.3 Interband Transitions

At this point I need to address one final subtlety that I have ignored until now. In all of the analyses presented so far I have assumed that the initial and excited electronic states were either both above the Dirac point or both below the Dirac point. However, there does exist the possibility of *interband* transitions wherein the initial state lies below the Dirac point and the excited state above, and vice versa. The only complication I must take care of is the fact that the chirality factor changes to $(1 - \hat{\mathbf{k}} \cdot \widehat{\mathbf{k} + \mathbf{q}})$. This reflects the fact that the spin chirality of the Dirac cone changes when passing through the Dirac point.

It turns out that this introduces two new, albeit very similar contributions to the integral. The first is an integral expression identical to Eq. (B.30) except with the upper and lower limits exchanged. Obviously for any given point in the space spanned by $\mathbf{k}$ and Ω only one of the two integrals will contribute since the other will have a lower limit whose value is higher than the upper limit, leading to a nonsensical integration. The second integral is an expression identical to (B.26) with the upper and lower limits exchanged. The same idea applies here as well.

Thus, with Eqs. (B.25), (B.26), (B.30), and (B.31) along with the matrix elements in Eq. (6.8), one can evaluate $\mathrm{Im}[\Sigma(\mathbf{k}, \omega, T)]$ numerically on the computer. With the imaginary part at hand a simple Kramers–Kronig transformation, also implemented numerically, will yield $\mathrm{Re}[\Sigma(\mathbf{k}, \omega, T)]$. This is sufficient to then compute the DFQ spectral function.

CPSIA information can be obtained
at www.ICGtesting.com
Printed in the USA
LVOW05*0151130617
537726LV00002BA/369/P

9 783319 447223